EPIGENETICS AND THE ARCHITECT

Epigenetics and the Architect

Evidence of Design at the Frontier of Biology

Thomas Woodward
and
James P. Gills, MD

Seattle   Discovery Institute Press   2026

Description

In 1953 Watson and Crick ignited the genetics revolution with their discovery of the double helix. Today a second revolution is underway thanks to the discovery of DNA's mysterious partner in the elegant waltz of cellular life—the epigenome. The sophistication of that dance is astonishing researchers and upending the genetic paradigm. *Epigenetics and the Architect* explores the latest in this unfolding revolution but goes a step further than other recent introductions to the subject. Evolutionary theory holds that all life diversified through a series of random mutations to DNA, but we now know that much of the information employed to build biological form is situated in epigenetic architecture beyond the reach of random genetic mutations and natural selection. As authors Thomas Woodward and James Gills show, this and the sheer intricacy of the genome/epigenome system are reinvigorating an ancient idea—that living things look designed because they are.

Copyright Notice

Library Cataloging Data

Epigenetics and the Architect: Evidence of Design at the Frontier of Biology
by Thomas Woodward and James P. Gills, MD
Cover design by Tri Widyatmaka.
144 pages, 6 x 9 inches
Library of Congress Control Number: 2026941708
ISBN: 978-1-63712-090-3 (paperback), 978-1-63712-092-7 (Kindle), 978-1-63712-091-0 (EPUB)
BISAC: SCI029000 SCIENCE/Life Sciences/Genetics & Genomics
BISAC: SCI027000 SCIENCE/Life Sciences/Evolution
BISAC: SCI049000 SCIENCE/Life Sciences/Molecular Biology

Publisher Information

Discovery Institute Press, 506 2nd Avenue, Suite 1700, Seattle, WA 98104
Internet: discovery.press
Published in the United States of America on acid-free paper.
First Edition, June 2026

ADVANCE PRAISE

In *Epigenetics and the Architect* Thomas Woodward and James Gills have written a masterful tour de force, providing a compelling account of the bewilderingly complex realm of the genome and epigenome. The epigenome is a complex integrated system of molecular controls that regulates which genes are expressed in particular cell types. The book provides a detailed and comprehensive account in easy-to-read, non-technical language of the ways the epigenome carries out this vital task. The authors describe how the attachment of small chemical groups to specific sites along the DNA molecule, or to specific histone sites in the chromatin, promotes or represses the expression of particular genes. And while the complexity of the epigenome, which regulates the expression of genes in one cell type, is in itself mind blowing, the total epigenomic complexity of any whole multicellular organism is vastly more complex, as each cell type has a different epigenome. And there is a further layer of complexity. As the authors explain, the vital controlled placement of the tags on different genes, through which downward control is imposed on gene expression and hence on the functional characteristics of different cells, is carried out by a host of mind-bogglingly complex molecular machines, all under the strict overall control of the cell. This remarkable book conveys more than any other book I have recently read something of the extraordinary, transcending complexity and wonder of biological systems, and every open-minded person who reads this book should agree that there is indeed here "Evidence of Design at the Frontier of Biology."
—**Michael Denton**, biochemist and author of *Evolution: A Theory in Crisis, Nature's Destiny,* and *The Miracle of the Cell*

The common view of the theory of evolution is that Charles Darwin proposed natural selection, in place of the crafting by a directed intelligence, to explain the variety among living organisms. As a physicist, the assertion that such unguided creation eliminated, by extension, the necessity of a creator anywhere in science is a triumphalism that was never justified. While Darwinism may explain the origin of species, the origin of the physical laws that allow the formation of DNA and the mutations that drive evolution remains unanswered. Additionally, Woodward and Gills point out that the popular narrative is incomplete and misleading even within biology. With epigenetics comes a control system that involves the entire cell and reveals the inadequacy of focusing on the genetic information in DNA without its expression. This more comprehensive picture not only enables cell variety based on the same genome but allows that random mutations in DNA are not the only way changes can be passed from one generation to the next. In *Epigenetics and the Architect*, we have a clear exposition of an important area of biology and a cautionary tale as well. Science is an "endless frontier," as Vannevar Bush famously put it, but only as long as we have minds unshackled by our prejudices.
—**Robert Kaita**, Senior Physicist, Princeton University

In accessible prose, Woodward and Gills show that the cell's information systems extend well past DNA, far into multiple layers of sophisticated, interdependent codes. This book is essential reading for anyone who wants to understand why shopworn Darwinism is crumbling.
—**Michael J. Behe**, Professor of Biological Science, Lehigh University, author of *Darwin's Black Box* and *Darwin Devolves*

Epigenetics and the Architect is a rather interesting book at several levels. Easily read in one sitting, it manages to review recent research into the epigenome, while showing both how it complements—and to some extent challenges—the dogmas of modern genetics and bolsters the case for intelligent design against a purposeless Neo-Darwinism.

But equally intriguing are the remarks sprinkled throughout the text about the role that lifestyle choices can make to enhance the workings of the epigenome, possibly across generations, suggesting a role for intelligent design theory in the practice of medicine.

—**Steve Fuller**, Professor of Social Epistemology, University of Warwick, UK, author of *Dissent over Descent: Intelligent Design's Challenge to Darwinism*

The molecular revolution in biology started in the 1950s with the discovery of the double helix structure of DNA, soon followed by the discovery of the genetic code and molecular machines for synthesizing RNA molecules and proteins. These hallmarks of design inspired the birth of the intelligent design movement in the 1980s and 1990s. Since then, research in molecular biology has revealed that the cell is even more advanced, with dozens of other codes and molecular machines that need to cooperate in an irreducibly complex way. Today we know that DNA is orchestrated and expressed through various epigenetic mechanisms in ways that guide embryological development and differ between cell types. Our epigenetics are affected by lifestyle, and to some extent those changes are passed on to subsequent generations. In short, the Central Dogma of molecular biology (DNA makes RNA makes proteins) has been replaced by the Cellular Dogma, with many codes throughout the cell. In *Epigenetics and the Architect*, Woodward and Gills describe this epigenetic revolution in a profound yet engaging way, accessible to a large audience. They convincingly argue that these newly discovered design features of the cell bear stronger evidence than ever for a Master Architect. I highly recommend *Epigenetics and the Architect*. It is my hope that the book is widely read and that anyone who reads it gets to know the Architect behind the designed cell.

—**Ola Hössjer**, Professor of Mathematical Statistics, Stockholm University, probability theorist with applications in population genetics, epidemiology, fine-tuning in biology, and the limits of evolutionary mechanisms; winner of the Gustafsson Prize in Mathematics

For anyone who was taught that all the information needed to build an organism is contained entirely in the DNA, this book is a must read to bring your science up-to-date. Our cells are permeated with vital information within, on top of, and beyond the DNA—and *Epigenetics and the Architect* will take readers on a highly educational yet readable and enjoyable tour of where that information resides throughout the cell. The complexity of the molecular machines and pathways regulating this epigenetic information is astounding—and it shouts "design" all throughout!
—**Casey Luskin**, co-author of *Science and Human Origins, Traipsing into Evolution: Intelligent Design and the Kitzmiller v. Dover Decision,* and *Discovering Intelligent Design*

I find the book *Epigenetics and the Architect* a comprehensive, interesting, and appropriate introduction to the intricacies of the interacting coding systems of the genetic makeup of organisms. It is comprehensive in compiling the latest advances in epigenetic studies by including various coding systems of the cell. It is interesting because it is written in layman's language and successfully employs science fiction storytelling techniques in places to explore real cellular functions. Its explanation of the various interacting coding systems of the cells and embryos leads to an inescapable implication of the design paradigm, which until the Enlightenment was the guiding principle of earlier scientific research. The authors then reiterate the arguments of design inferences, irreducible complexities, and the inability of undirected evolutionary processes by natural selection to produce novel protein functions or body plans. They also include evidence that theistic faith tends to produce behaviors that maintain or improve epigenetic health, which can be passed on to future generations. The book makes a strong contribution to the arguments that the design hypothesis is a viable and productive venue for future scientific research.
—**Pattle P. Pun**, Professor of Biology Emeritus, Wheaton College

Are you ready to experience a "fantastic voyage," one that takes you right inside the cell to update you on the latest scientific discoveries on how it actually works? It's heady stuff, but there's no need to feel intimidated. In *Epigenetics and the Architect*, Thomas Woodward and Dr. James P. Gills do a masterful job of safely piloting you from the old dogma of the genome coding only for RNA and proteins, and rife with "junk," to the revelation that it's the epigenome that controls DNA expression, through non-junky, non-coding DNA and many other non-genomic mechanisms. Why does all this matter? Besides showing that evolutionary biologists missed the boat on accurately predicting how life actually works, the book's revelations also explain how the cells in your body can have virtually identical genomes but differentiate into hundreds of different cell types. It also opens up new vistas into healthcare and, as the authors opine, "adds immeasurably to one's appreciation for the beauty and ingenuity of life's design." Be amazed, be elevated, and be humbled! Highly recommended for the layman who wants to be on the forefront of scientific knowledge.
—**Howard Glicksman**, physician and co-author of *Your Designed Body*

The compelling book *Epigenetics and the Architect* by authors Woodward and Gills deserves special recognition for two reasons: It addresses highly complex genetic topics using cutting-edge scientific information while remaining accessible and engaging for non-specialist readers, and it demonstrates that the extraordinary complexity of the cell's molecular machinery—along with its genetic and epigenetic regulation—provides strong support for the theory of intelligent design. Woodward and Gills ask whether such remarkable molecular complexity and epigenetic regulation could have arisen with purpose or without purpose. The answer they provide, supported by multiple and compelling lines of evidence, is that this astonishing complexity could only have emerged with a clearly defined purpose. It is an inference to the best explanation. Chance and necessity, proposed by Jacques Monod in the 1970s as the driving forces behind biological complexity, are now revealed as absurd in light of these new discoveries. The materialist philosophy underlying Darwinian evolution and the modern synthesis has become obsolete and reaches its limits in the face of the findings of epigenetics. This highly recommended and groundbreaking book will lead readers to discover the Architect behind the marvelous design of molecular genetics and its epigenetic regulation.

—**Ricardo Bravo**, Professor of Zoology and Dean of the Faculty of Oceanography and Natural Resources, Universidad de Valparaiso, Chile

If Darwin had known more about biology, he never would have proposed his theory of evolution. Here Woodward and Gills present such information, in the epigenetics category, with wonderful descriptions and stories about the science, the scientists, and the biological revelations they are uncovering. Highly recommended.

—**Cornelius G. Hunter**, biophysicist and author of *Science's Blind Spot: The Unseen Religion of Scientific Naturalism* and *Darwin's God: Evolution and the Problem of Evil*

CONTENTS

1. DNA and Beyond

ONE OF THE MOST AMBITIOUS AND PROMISING SCIENTIFIC PROJECTS of the last half century was the Human Genome Project, launched in 1990. It aimed to map the entire human genome, down to the exact sequence of DNA's four-character alphabet of chemical letters arrayed along the double helix. This herculean goal, analogous to landing a man on the moon, took billions of dollars and a decade-plus of labor by thousands of scientists.

The first milestone was reached in 2000. President Bill Clinton called a news conference to announce that a rough draft of the genome had been assembled. After more refining and cross-checking of the data, the project was completed in 2003 and a final draft of the genome was published. Scientists had delivered a complete map of our DNA, right down to the positions and sequences of tens of thousands of genes on the different chromosomes, with each protein-coding gene containing a wealth of digital information. (Microsoft founder Bill Gates commented that "DNA is like a computer program but far, far more advanced than any software ever created."[1]) Thanks to the coordinated efforts of geneticists around the globe, anyone could access online the exact spelling of the entire 3.1-billion-letter DNA database for *Homo sapiens*.[2]

The project was a magnificent success, and hopes were running high for its practical benefits. At the June 2000 news conference, President Clinton gushed, "With this profound new knowledge, humankind is on the verge of gaining immense new power to heal. Genome science will have a real impact on all our lives and even more on the lives of our children. It will revolutionize the diagnosis, prevention, and treatment of most, if not all, human diseases."[3] Nor was it just politicians expressing such high hopes. Around the same time, Mike Stratton, of the Cancer Genome Project, remarked, "It would surprise me enormously if in twenty years the treatment of cancer had not been transformed." And UK Science Minister Lord David Sainsbury exclaimed, "We now have the possibility of achieving all we ever hoped for from medicine."[4]

Alas, many of those expectations remain unfulfilled. For example, when the project was being organized, there were grand expectations that it would quickly lead to a host of powerful gene therapies. That hasn't happened. As the headline of a front-page article in *The New York Times* announced, "A Decade Later, Genetic Map Yields Few New Cures." In the article, writer Nicholas Wade expresses disappointment that

> medicine has yet to see any large part of the promised benefits. For biologists, the genome has yielded one insightful surprise after another. But the primary goal of the $3 billion Human Genome Project—to ferret out the genetic roots of common diseases like cancer and Alzheimer's and then generate treatments—remains largely elusive. Indeed, after... years of effort, geneticists are almost back to square one in knowing where to look for the roots of common disease.[5]

The Human Genome Project was an undeniable success in advancing our knowledge of our DNA. So, how could the conquest of such an important frontier in genetics produce such underwhelming results in terms of medical breakthroughs?

One thing that held things back, we would suggest, was insufficient progress in a genome-adjacent frontier—a situation, however, that is beginning to change. Scientists are learning more and more about a second biological encyclopedia of information that resides above the information stored in DNA. Researchers are decoding a system in the cell—a sophisticated software network situated beyond DNA—crucial to the differentiation of a single, fertilized egg cell into hundreds to thousands of cell types during human development.

This higher control system also appears to be a factor in aging processes, cancer, and many other diseases. It guides the expression of DNA, telling different kinds of cells to use different genes, and to use them in the precise ways that meet the needs of those different cells. This "information beyond DNA" plays a crucial role in each of our thirty trillion cells (give or take a few trillion), telling the genes when, where, and how they are to be expressed.[6] This is biology's mysterious new frontier—the epigenome.

None of this is to say that progress in genetics has slowed. Practically every month sees researchers publishing new discoveries about DNA and its information storage and processing abilities, along with fresh insights into DNA's companion molecules, RNA and proteins, built from the information stored in DNA. But interwoven with these stories are reports of scientists discovering fresh details about the control system that sits above our genetic riches, the many byways of the wondrous epigenome now being diligently mapped and cataloged.[7]

Much of this epigenetic architecture resides very close to our genes, dynamically connected to the double helix in the form of a multi-layered system of tiny chemical tags and molecular structures such as the nucleosomes—tiny spools coiled with DNA. (See Figure 1.1.) Unraveling the details of this extraordinary interplay of genome and epigenome is proving enormously fruitful. It is also raising fresh questions as to how such a sophisticated information-processing system could have arisen in the history of life.

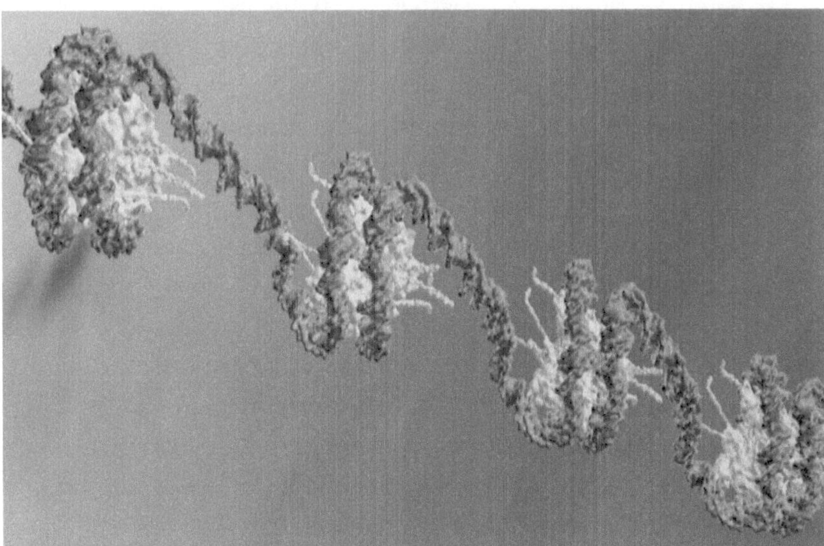

Figure 1.1. A stretch of DNA wrapped around spools made of a tightly packed cluster of histone proteins (the light colored globular molecules). Eight histone proteins combine together and are wrapped with DNA to form a nucleosome. Here, four nucleosomes are visible, with tails on the spools.

The term *epigenome* commonly refers to the various chemical tags that attach to histones and nucleotides, which together direct various epigenetic regulatory mechanisms. Yet, in our journey through the epigenome, we will use the word in a slightly broader sense, to include all layers and levels of cell memory and stored information found beyond DNA and RNA—all the way out to our cellular membranes, which have their own sophisticated pockets of information. This is the realm of inheritance not rooted in our genes.

Picture a human genome, composed of DNA, as a sailing ship docked in calm waters. When the winds rise and the captain wants to venture out to sea, he sets the sails to catch the wind and moves the rudder to direct the ship toward its destination. In our analogy, the captain, the sails, and the rudder are the multilayer epigenome. We want to explore and understand every part of this biological ship at sea, both the DNA and every dimension of the cell's epigenetic programming that directs the expression of DNA.

Junk Gets a Promotion

DNA, often described as the master code of life, is flashing an impish smile. She has been coy of late, harboring shocking scientific secrets. During the past three decades, this thread-like, spiral molecule has played a game of genetic hide-and-seek with scientists. Fortunately, she has left many helpful clues for us to follow, and researchers have been stunned by what these clues are telling us.

One is the discovery of a sophisticated "splicing code" embedded within the familiar DNA sequence.[8] This set of instructions enables a single gene to be knit together into a host of gene products—the messenger RNA "gene copies" that serve as blueprints for building proteins. The variety of such products from one gene is shocking, "numbering in the hundreds and even thousands," as Douglas Black explains in the journal *Cell*.[9] It is like a brilliant chef producing a single "super recipe" from whose instructions cooks can produce a thousand different sumptuous dishes.

Perhaps the biggest headlines came from a pair of scientific discoveries, compliments of the combined effort of dozens of laboratories across the world, known as the ENCODE Project. Berkley Gryder, head of the Gryder Lab at Case Western Reserve University, described the findings as an "earthquake" in the molecular biological community.[10] This global study turned the prevailing picture of DNA on its head. Previously, vast stretches of the human genome had been deemed "junk DNA." Estimates have varied, but for decades molecular biologists frequently predicted and believed that 90 percent of the genome—even as much as 95–99 percent—was composed of useless genomic gibberish.[11] According to this older view, these stretches of DNA (unlike our protein-coding genes) were debris accumulated during eons of evolution, precisely what one would expect from an evolutionary process devoid of foresight and planning and fueled by a long series of random genetic mutations.

Then ENCODE published its findings, revealing that in many cases the mysterious "junk" sequences are read and copied, and the copies are used in numerous cellular functions. They are anything but junk. Ewan Birney, a pioneer in the field of bioinformatics, estimated that at least 80 percent of our DNA has biological function, and added, "It's likely that 80 percent will go to 100 percent. We don't really have any large chunks of redundant DNA. This metaphor of junk isn't that useful."[12]

By 2024, three scientists completed a review and found over 800 scientific papers describing the functions of formerly labeled junk DNA.[13] Much remains to be learned, but one thing is already clear: The credibility of the junk DNA framework has been fatally damaged. Textbooks are being rewritten to accommodate this surprising reversal.

Also of note, many of our genes have been found to code for various RNA copies that are never used for protein production. For instance, one type of RNA, called *microRNA*, is tiny, averaging twenty-two nucleotides long. Discovered by biologists Victor Ambros and Gary Ruvkun, microRNAs (or "miRNAs") are crucial to regulating growth and development, as well as gene expression and metabolism. Their importance was underscored by the awarding of the 2024 Nobel Prize in Medicine to this duo.[14]

This miniature RNA is just one of many RNA molecules that have been discovered. Intriguingly, many of these RNAs don't leave the nucleus. They spring into action inside the nucleus, where they carry out vital functions. Such findings challenge the traditional view of our genetic blueprint as a tidy collection of independent genes, and instead point to a complex network in which genes, along with regulatory elements and other types of DNA sequences that do not code for proteins, interact in ways not yet fully understood.[15]

The Birth of a Discipline: Epigenetics

The other big discovery is the epigenome, DNA's partner in the elegant waltz of cellular life, which determines how genes are expressed. Or to borrow another metaphor from the world of music, it's not unlike the way a skilled conductor guides an orchestra. The study of this complex system is called epigenomics or, more commonly, epigenetics. Figure 1.2 compares genetics with epigenetics. All over the world researchers are probing the mysteries of the epigenome, and the complexity of the system coming into focus seems to grow with each passing month.

We have learned much about the clever architecture that enables this system to work so efficiently. Many of its functions are directed by a chemical software program with its own set of codes, connected by tiny signals and switches.

Two of these epigenetic codes are embedded close to a cell's DNA in a double-tiered library of instructions. This dual library differs from cell type to cell type so that, for example, a brain cell's epigenomic information is noticeably different from that of a muscle cell. According to Gryder, if various developmental stages of cells are included, the number of distinct epigenomes might tally in the many thousands.[16]

When we zoom in on the intricate nooks of DNA's molecular landscape, we see millions of these chemical switches. Some epigenetic signals are hard to spot—tiny molecular tags placed on the double helix itself. A second epigenetic code involves various distinct markers attached to the spools DNA is coiled onto, the histone proteins pictured in Figure 1.1. The wrapped DNA, plus the histone-based spools as well as other proteins that help package DNA into chromosomes, are together called chromatin. Chromatin makes up the stuff of our forty-six chromosomes, coiled together deep inside the nucleus. Intriguingly, chromatin has been found to regularly undergo purposeful transformations, known as chromatin remodeling. These shape-shifting changes are a prime focus of epigenetic research.

GENETICS VS. EPIGENETICS

Field of Study	Genetics	Epigenetics
Complete Library	Genome	Epigenome
Key Functions	Code for RNA and Proteins	Controls DNA Expression
Information Format(s)	DNA's four-character alphabet (A, T, C, G) used in gene sequences and other formats	Chemical tags, such as acetyls, on histone tails; methyl groups on DNA;[17] and other molecular patterns all over the cell
Variation Among Cell Types in an Organism	Very little: The genome is virtually identical in all nucleated cell types	Much: Hundreds of cell types and epigenomes—perhaps thousands if transitional states included
Heritable Changes?	Germ-cell mutations are inherited	Modification can be passed on to subsequent generations
Changes by Lifestyle?	No	Yes: many

Figure 1.2. Genetics and epigenetics compared.

Emerging evidence points to additional layers of subtly coded information built into other parts of the cell, including in the cell membrane and the interior structural members of a cell. This strange new realm of functional information, written into parts of our cells distant from DNA, can be startling when one hears of it for the first time. It's a bit like discovering that the digital memory in a computer is not confined to the hard drive. Imagine being told that millions of bits of vital data are inscribed in other specialized codes embedded in the interior wiring, keyboard, screen, outer casing, and many other parts of the computer.

Reprogramming Family Health for Generations

Perhaps the most sobering discovery emerging from this research field is that our system of epigenetic control can be modified by our lifestyles, and these changes passed down to our children, grandchildren, and possibly beyond.[18]

For example, a study by Lars Bygren of the Karolinska Institute in Stockholm focused on the health histories of ninety-nine families in a tiny village in a remote agricultural region called Norrbotten, in northernmost Sweden.[19] As Bygren studied the life patterns in this village, where his own father had grown up, he uncovered a stark reality. A pattern of binge-eating during a period of abundant harvests seemed to have reprogrammed the epigenetic system of young boys in the village, dealing a devastating blow that lasted for many decades. By studying the patterns of diet and longevity in these lineages, Bygren concluded that lifespans of the next two generations of farmers were shortened appreciably because of a single year of gluttony.

These findings, originally reported in prestigious science journals, was highlighted in a *Time* cover story, "Why Your DNA Isn't Your Destiny," which surveyed the explosion of epigenetics research. "Bygren… showed that the grandsons of Overkalix boys who had overeaten died an average of six years earlier than the grandsons of those who had endured a poor harvest," reported journalist John Cloud. "Once Bygren and his team controlled for certain socioeconomic variations, the difference in longevity jumped to an astonishing 32 years."[20]

These discoveries have brought us to the edge of a revolution in the biology of inheritance. Several lines of evidence suggest that patterns of daily living—including diet, stress, smoking, and exercise—may have the power to reprogram one's epigenetic system and that of future offspring. How deep do such epigenetic changes go—for good or ill? How many generations may reap the effects of these epigenetic alterations? Answers thus far are hazy, and the subject remains an area of ongoing research. But what is clear is that our epigenome is somewhat malleable, and changes can be passed on at least as far as one's grandchildren. Cloud concludes the *Time* article thus: "It will take geneticists and ethicists many years to work out all the implications, but be assured: the age of epigenetics has arrived."

A Multilayered Revolution

In sum, scientists continue to probe the incredible genome, even as they plumb the mysterious epigenome. These advances are worthwhile for their own sake—for the intrinsic value in deepening our understanding and appreciation of the intricacies of organic life. But this new double focus—genetics + epigenetics—also may teach us how to modify the epigenome through careful changes to diet, exercise, and other life patterns so as to improve our health[21] and perhaps even the health of our descendants.

As we sketch a picture of this molecular architecture, we will highlight recent findings about DNA and show how the epigenome's switches and gadgets work thanks to complex biological machinery. The machine descriptor in this context, mind you, must be taken with a grain of salt. The term is meant to convey the fact that each of these molecular devices is marked by a purposeful arrangement of precisely tailored parts into an organized whole, one that achieves a function that no mere jumble of parts would. But the machine imagery provides only a partial picture. It's not that it exaggerates the situation. Just the opposite. It risks diminishing the advanced nature of these marvels, devices that possess an intrinsic and goal-directed purposiveness, often including the ability to self-assemble and self-repair. These twin powers are absent from the relatively crude technology devised by our most advanced human engineers.

We persist, however, in using the language of machines because we are all familiar with various manmade machines to one degree or another, and the language serves well to shatter the common misconception of microbial life as relatively crude and simple beside cutting-edge human technology. Take the most advanced human factory you can think of, and now imagine that it also can self-replicate, self-repair, and run without human factory workers; then you will begin to get an idea of the sort of "machines" we are talking about in microbiology.

As we explore this frontier of discovery, we will take a couple of trips via an imaginary micro-miniaturized submarine into the realm of the cell and visit its DNA-packed nucleus. The submarine is ficti-tious, of course, but what its crew observes are the very real discoveries made by advanced laboratories around the world.

Purposeless or Purposeful?

Because we are dealing with foundational scientific discoveries, we will address several "So what?" questions in the course of the book. One, already noted, relates to health and wellness. Another is whether these findings strengthen the impression of an inherent purpose to life and the natural world.

Moving deeper, we will consider how these findings challenge our worldview—our *Weltanschauung*, to use the German term. After exploring the new vistas of integrated, multilevel cellular complexity, we will ask, Who or what designed and built this system? Was it a purely blind, material process, as modern evolutionary theory holds? Or is there evidence that this sophisticated architecture sprang from some designing intelligence—from an Architect? This question will be tackled using an approach long employed in the historical sciences, called the method of multiple competing hypotheses, or inference to the best explanation, an approach elucidated with particular rigor by the late Cambridge philosopher of science Peter Lipton.[22]

We approach with special care this issue of how cellular complex-ity arose, aware that modern evolutionary theory is closely associated in many people's minds with cherished values such as scientific en-lightenment, critical reasoning, and educational progress. As a result, when one points out empirical problems with its story of how life's interconnected complexities arose, one reaction may be scandalized astonishment mixed with incredulity. This sense of shock is all the more likely if one has not followed the twists and turns in this scien-tific discussion in recent years. Those who have not will be unaware, for instance, that in November 2016, the Royal Society, the oldest and among the most distinguished scientific societies in the world, met in London to discuss questions in biology that remain unanswered by the neo-Darwinian synthesis. At the gathering, several leading life scientists revealed their deep dissatisfaction with textbook orthodoxy regarding how new life forms arose.[23]

Another notable moment occurred in 2022 with the publication of an article entitled, "Do We Need a New Theory of Evolution?" Appearing in a June issue of the UK's left-leaning *The Guardian*, this controversial survey by science writer Stephen Buranyi featured the heretical ideas of various biologists from what is known as the Third Way of Evolution, or the Extended Evolutionary Synthesis (EES).[24] This group relegates the neo-Darwinian mechanism to a secondary role in explaining the emergence of novel biological form, focusing instead on new mechanisms they hope will shore up the weakened center post of evolutionary theory.[25]

Meanwhile, various recent studies have expanded the empirical resources of those championing the design hypothesis.[26] One book-length treatment focuses on empirical evidence of genetic mutations in action, and on the nature of those mutations that have provided notable examples of evolution. The book, *Darwin Devolves* (2019) by Lehigh University biologist Michael Behe, reviews a series of revelations about cases of microevolution, such as beak variations in Galapagos finches and the rise of polar bears from brown bears. These and many other transitions, it turns out, were enabled not by the production of new genes or new molecular biological machines. Rather, the new research shows that these cases arose by the breaking or blunting of existing genes. Behe, it should be noted, was not reporting on his own laboratory research. He was assembling data from a wide array of peer-reviewed research published in mainstream science journals.

Having synthesized this information, Behe concludes that the evidence points away from any fundamentally novel genes arising in creaturely DNA through a series of random genetic mutations, with or without natural selection. The uncomfortable question then arises:

How did wholly new genes, with their tens of thousands of base pairs of software-like code, arise in an animal's genome in the first place, not to mention the sophisticated epigenetic control system essential to gene function?

In a book review in the journal *Science*, critics tossed verbal grenades at Behe's devolution argument. Behe provided a detailed response. Its opening paragraphs provide a compact summary of five ways that, according to him, the attack on *Darwin Devolves* falls short. He ends his lengthy rebuttal of the book review thus:

> Can any important conclusion be drawn from this train wreck of a review? Consider that Richard Lenski is perhaps the most qualified scientist in the world to review the argument of *Darwin Devolves*. Lenski has spent decades overseeing the most extensive, most acclaimed laboratory evolution experiment conducted to date, for which he was elected a member of the National Academy of Sciences. His own work is a major focus of *Darwin Devolves*. He could easily and casually have pointed out any problems with the argument all by himself, merely by wielding Darwin's putatively powerful theory and his own expertise. Yet he and his co-authors spend the entire review deriding the author, barely mentioning Lenski's own work, recounting old criticisms by other people, and leaning heavily on aged theoretical conjectures instead of new experimental results.
>
> The implication is clear. Just as Russell Doolittle unwittingly showed, simply by his mistaken citation, that no [Darwinian] explanation for the origin of the blood clotting cascade existed, so the reviewers show that there is no answer to the problems for unguided evolution described in *Darwin Devolves*. Although it serves a useful role in understanding changes at the margins of biology, as an explanation for the overarching structure of life Darwin's theory is defunct.[27]

As more questions arise about modern evolutionary theory, more scientists than ever, all around the world, are asking if textbook explanations still hold water. Is it plausible that mindless, undirected natural processes crafted all the cell's high-tech hardware and software?

Such a claim would seem, at the very least, to be under considerable stress from the weight of new data.

Charles Darwin himself had no way to glimpse the intricacies of the cell. He and his contemporaries viewed the cell as a fairly simple substance. But when one advances from Darwin's Victorian England to the twenty-first century, we find a significant change in perspective. In recent years, as biologists and geneticists have continued to unravel the cell's mysteries, phrases such as "staggering complexity" and "infinitely complex" have appeared in the literature.[28] The sheer sophistication of the nano-world opening up before us has repeatedly shocked the research community. DNA is more information-rich than we imagined, and we now know that it is tethered to a mysterious control system that directs the molecular orchestra.

Our hope is that the vista upon vista of fantastic microscopic architecture we tour in these pages will inspire not only a deeper understanding of molecular biology but also a renewed sense of wonder and a measure of humble curiosity about what we have yet to learn. In a speech at Rice University in 1962, President John F. Kennedy said, "The greater our knowledge increases, the more our ignorance unfolds."[29] This truth manifestly applies to the life sciences. Biologists are probing ever deeper into the unknown. In doing so, we not only find out new truths, but also discover new vistas of ignorance. In the case of biological information, that dance of genetics and epigenetics, each new discovery raises fresh questions and opportunities to realize just how much we do not know, and how much more there is yet to discover.

2. What Darwin Didn't Know About the Cell

C HARLES DARWIN WAS UNDENIABLY A BRILLIANT MAN OF SCIENCE. But he was also undeniably a man of his time, meaning he propounded his theory of evolution on a foundation of almost total ignorance about the nature of the cell. In 1859, when his work *On the Origin of Species* first appeared, the series of discoveries that would begin to unfold the cell's marvelous complexity lay all in the future. Thus, Darwin knew nothing about the details of genetic inheritance carried within cells—not even Gregor Mendel's momentous discovery of dominant and recessive genes, published in 1866 but unnoticed until after Darwin's death.

Nineteenth-century biology's unawareness of cellular complexity does not, by itself, deliver a verdict on Darwin's theory; but clearly it did serve to remove a serious obstacle to the theory's postulation of nature-driven incremental change. In Darwin's mind, any new bodily structure could be developed step by tiny step, over eons of time, beginning with the first primitive life, which he and his fellow naturalists assumed had been unsophisticated in the extreme, something on the order of protoplasmic goo in a sack. This misapprehension about the cellular realm meant that they weren't required to confront the cell's advanced technology and powerful appearance of design, including its system of inheritance coursing with sophisticated digital code. The Darwinian era's ignorance of all this brilliant engineering—so suggestive of foresight and planning—raised the relative prospects for his proposal, namely, that nature's unguided, unintelligent forces had the power to create the world of biology.[1]

Figure 2.1. Englishman Charles Darwin, author of *The Origin of Species*, released in November 1859.

Some of our readers are already well apprised of the main findings of molecular biology in the latter half of the twentieth century—the double helical structure of DNA, the form and functions of various RNAs, the genetic code, and how proteins are formed and go on to perform a host of duties like so many specialized robots in a factory. But since this is a book for general readers, and many of us have forgotten more of our high school and college biology than we learned, a short overview is in order.

We often talk loosely about "complex life" and "simple life," with the latter referring to microbes such as bacteria. But as we saw, even single-celled organisms are complex—mind-bendingly so. Consider the tiny *E. coli*. You would need to line up ten thousand of them end to end for them to stretch a single centimeter. Yet each one of these bacteria contains more digital information than a 500-page novel; and all of that biological information is orchestrated in a way that allows the bacterium to function as a microscopic robot that can obtain its own fuel, repair itself, and self-reproduce—a robot and a robot factory in one.

If you think of this tiny organism's genetic database in terms of computer files, it comes to a megabyte of information, with over four thousand separate files, better known as genes. The human genome, for comparison, is about a thousand times larger. Inside each human nucleus are two copies of the human genome, each with more than 3.1 billion letter pairs. This is the information equivalent of thousands of books—that much in each cell nucleus. (Human DNA is packed into a spherical nucleus, while the DNA of *E. coli* and other bacteria is woven in the shape of a loop.)

The inside of a cell roughly resembles a 3D maze of cables and tubes stretching in every direction. The cell's DNA takes the form of a double helix, like a twisting ladder, except it's mostly coiled up like hundreds of tiny kinks in a rope. All the DNA information together is referred to as the genome. Every creature on earth—whether microbe, plant, or animal—possesses a unique genome functioning as a storehouse of precisely written instructions for that particular creature.

Each rung on DNA's twisting ladder is formed from a pair of chemical units called nucleotide bases. These units function as alphabetic characters in a code. DNA is not merely *like* a code. DNA *is* a code. (See Figure 2.2.)

English has a twenty-six character alphabet. DNA has a four-character alphabet, represented by the letters A, T, C, and G, referred to as the four bases. These four letters stand for the four distinct chemicals of this genetic alphabet. The letter A stands for adenine, T for thymine, C for cytosine, and G for guanine. A single rung in the twisting ladder of DNA has a pair of chemical letters, either AT or CG, called a base pair. A segment of DNA with twenty-one rungs, then, would have twenty-one base pairs.[2] (See Figure 2.2.)

As we will see in more detail below, DNA's language is written not only with an alphabet of four letters, but also in a format of three-letter words. Thus a segment with twenty-one rungs will spell out seven genetic words of three letters each, called *codons*. This language format is also used in DNA's sister molecule, RNA, which is a copy of DNA. Each three letter-codon can function in one of two ways. First, a few codons act as signals, either for the start of a protein chain (like a capital letter initiating a sentence) or to signal the end of the protein chain (like a period marking the end of a sentence). Those are called "start" and "stop" codons respectively. Second, most codons—well over 90 percent—act as special code words, with each specifying that a particular amino acid be added to the growing protein chain. The start codon (AUG), we should note, also codes for an amino acid (methionine). The stop codons, in contrast, do not normally specify an amino acid.

A short gene may only have several dozen codons, while longer genes can tally many thousands of codons. How many genes are found in the DNA of a single bacterial cell? An *E. coli* cell, for example, can harbor well over five thousand distinct genes, and a typical gene may contain a chain of a thousand DNA base pairs or more.[3] And all that information, vital to the many functions of the cell, is so tightly condensed that it fits into a space thousands of times smaller than the period at the end of a sentence. Scientists have concluded that no entity in the world stores and processes information as efficiently as the DNA molecule.[4]

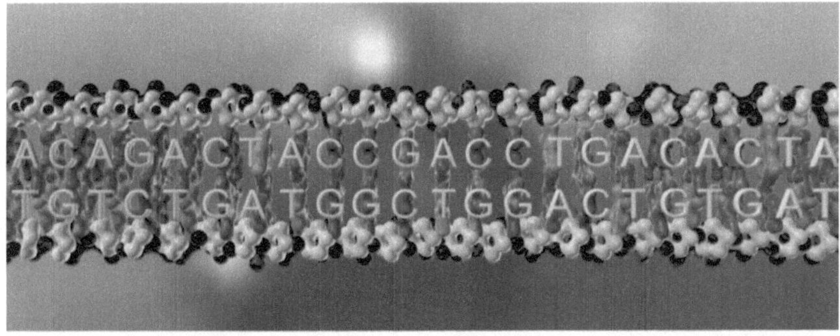

Figure 2.2. DNA rungs are shown in a flat format, making visible the pairings of A with T and C with G. The outer edge shows deoxyribose sugars alternating with phosphate molecules.

DNA, with its iconic double helical structure, is the most famous of the many molecules inhabiting a cell, but it can't do its work alone. It depends on two types of partners—RNAs and proteins. Together with DNA, these are often described as informational macromolecules.

Let's unpack this. The term macromolecule fits because each of these three molecules forms a long chain, typically containing many hundreds or thousands of smaller molecules linked together and functioning as letters in an information system. The term "informational macromolecules" accurately describes the richness and density of letter-by-letter coded information in those long chains; the exact sequences of those letters spell out important messages.

RNA is reminiscent of the twisting DNA ladder, except it's as if DNA was a chemical zipper that was unzipped. Or imagine someone cutting the DNA ladder down the middle and setting one half aside. (Proteins, in contrast, come in many shapes and sizes, but many resemble a long, folded-up chain of lumpy pillows.)

DNA's four-character alphabet (A, T, C, and G) is used to generate the twenty-character alphabet of amino acids. And that alphabet of twenty distinct amino acids codes for all the various protein machines in your body and in the body of every living organism. RNA is the link between DNA and the proteins DNA codes for.

Just as the DNA rungs contain coded information, so too do the stubby half-rungs sticking out from RNA's long side chain. Each half-rung is one letter from RNA's specialized four-letter alphabet.

In the RNA alphabet, one of the letters in DNA, the thymine (T), is replaced by *uracil*, a chemical represented by the letter U. RNA pairs A with U, and C with G.

RNA is every bit as indispensable as DNA. We depend on the trillions upon trillions of precisely coded RNA chains in our bodies. Their best-known role is transferring digital copies of key portions of DNA. But we're discovering other vital roles RNA plays, especially inside the cell nucleus. RNA genes play a special role in gene expression, setting up key differences between species. Every year, it seems, new important RNA molecules are discovered.

Normally the DNA double helix is wound tightly into bundles within the cell nucleus. The nucleus, recall, is crammed with DNA, thousands and thousands of precise sequences; and about half the sequences are aperiodic and specified,[5] as information theorists put it. That is, they are not simple repeating patterns like atcg-atcg-atcg. Nor are the sequences random gibberish. Instead, they are like the sequences in software or a book.

The cell nucleus is a compartment within the larger container of the cell. Every species of organism on earth—aside from bacteria and archaea—stores most of its DNA in a nucleus. If we were to zoom in on a nucleus's thin wall, we would find it perforated by tiny holes called nuclear pores. These are not simple holes, however. They are high-tech, mechanized gateways, with each pore using at least thirty interlocking machine-like parts. In a typical human cell, the nucleus has about two thousand pores scattered over its surface.[6]

Now let's turn to the high-tech aspects of a single gene. A gene typically also has a control panel, a somewhat shorter sequence of the double helix and not part of the protein-coding DNA sequence itself. Instead, it's a regulatory region that proteins and other chemicals bind to. Some parts of it promote the process of transcribing the DNA into RNA; others inhibit it.[7]

To copy the DNA information, the twisted DNA ladder opens, splitting like a zipper and exposing the individual DNA letters. Inside a nucleus, this process is orchestrated by sophisticated protein machines, each performing a specialized function. In the process the DNA's crucial information—the building plan for a protein molecule—is safely transferred to the RNA copy. Online animations of this molecular dance give the powerful impression that cells are run by tiny, advanced droids, a factory team that puts our most advanced robotized factories to shame.

In the cells of animals and plants, something remarkable happens next. The RNA chains, right after they are built, are not yet ready to code for the specific protein needed there. The initially transcribed RNA segment contains the potential for many different kinds of proteins, and now it becomes the target of a highly competent splicing and editing process aimed at tailoring the RNA segment to produce the particular kind of protein needed at that time and place. Molecular robots appear and swoop down on the RNA, adroitly performing the needed snip-snip so that what needs removing is removed.[8] After the RNA chains are assembled and properly edited, they slip out of the nucleus and are ushered to a group of strange-looking RNA-and-protein machines known as ribosomes, which perform the vital work of turning RNA into protein machines.

Ribosomes are much smaller than the cell nucleus but hefty compared to, say, nucleotides or amino acids. Each ribosome is made of several gigantic RNA molecules and a team of more than fifty proteins. Altogether, each ribosome has more than 300,000 atoms. There are about 20,000 ribosomes in a typical bacterial cell, such as *E.coli*, though there can be as many as 72,000. In some mammalian cells—liver cells, for example—there can be an army of several million.

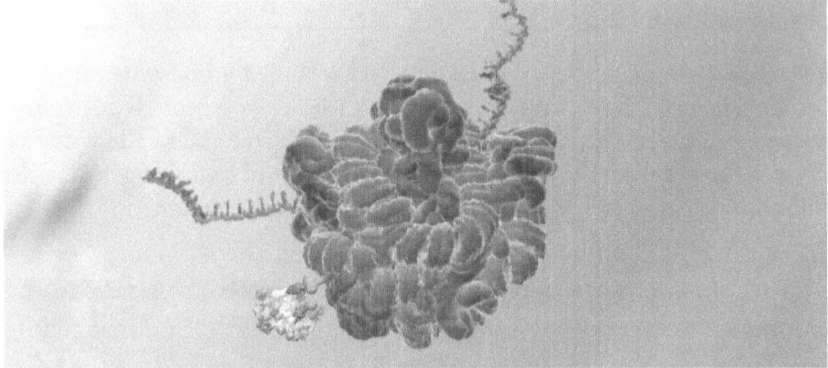

Figure 2.3. The ribosome is a crucial machine in cells. It is the central factory for translating strips of mRNA code into functioning proteins.

To build a given protein, this incredible machine uses the pattern it receives from the RNA chain that has exited the nucleus. This chain, called messenger RNA, enters a tunnel on the side of the ribosome machine, which takes the tiny RNA strip of information and translates it into a new language—the protein language with its twenty-character amino acid alphabet. As the RNA is fed into the ribosome machine, it passes through it like a train through a tunnel. In the process, the RNA is read, codon by codon, three letters at a time. Each codon is translated into the correct letter from the amino acid alphabet. And from a lengthening chain of these amino acids, the ribosome produces the new protein chain.

Figure 2.4. In this artist's rendering of the inside of a ribosome tunnel, a messenger RNA half-ladder moves along while a series of tRNA cross-shaped molecules, each holding an amino acid, match their code with the correct three-letter RNA codons. Result: a chain of amino acids is built, one amino acid at a time, which folds into a functioning protein.

In animations of this process, a chain emerges from one end of the ribosome, pictured something like a plump, headless animal that is rapidly growing itself a long fluffy tail. As the tail exits the ribosome, it begins curling together, folding into a lumpy mass. And as the last words of the RNA strip pass through the ribosome's tunnel and are read and translated, the final RNA word commands, *Stop*. Then the protein string has been born, and it detaches from the ribosome.

The translation process is akin to how a smartphone reads patterns on its stored memory and turns them into a ringtone or entire song. In this analogy, a super-short protein of fifty amino acids is like the ringtone. And the longest known protein—the muscle protein called titin, with about thirty thousand amino acids—is more like, say, Beethoven's Ninth Symphony.

As the protein is assembled by the ribosome, it begins to curl and fold in on itself. Folding also continues after translation, often in another cellular organelle called the endoplasmic reticulum. During the folding process, a protein is compacted into a complex shape a fraction of its original size. And compacted not like a sheet of paper wadded up at random; the compacted shape is a precisely preprogrammed three-dimensional fold determined by the chemical properties of the amino acids in the protein chain,[9] the particulars of which are vital to its proper folding, and to fulfilling its critical functions. A protein's specific 3D shape depends on the exact sequence of its letters—the amino acids. The thousands of different kinds of proteins are built from thousands of different amino acid sequences, each yielding a distinct folded shape—from simple ring structures and tubular shapes all the way up to our favorite, a robotic worker called kinesin.

The kinesin's body structure involves four proteins joined into one unit that strides along cellular highways called microtubules. Using its pair of protein appendages, which look like skinny legs with feet, it hustles along at upwards of a hundred steps per second. Scientifically accurate animations show this industrious delivery worker hurrying down a tube-shaped highway in the cell, pulling a bulky spherical mass attached to its shoulders,[10] a cargo that the worker bot will deliver to just the right location.

Kinesin is but one of many advanced molecular machines in the cell. Even the simplest cells, far from the crude affairs they were once believed to be, are brimming with biological information and buzzing with ingenious robots. Australian geneticist Michael Denton memorably summarized the reality: "Although the tiniest bacterial cells are incredibly small… each is in effect a veritable micro-miniaturized factory containing thousands of exquisitely designed pieces of intricate molecular machinery, made up altogether of one hundred thousand million atoms, far more complicated than any machinery built by man and absolutely without parallel in the non-living world."[11]

Figure 2.5. The kinesin molecular machine at work.

The Cell's High Technology—Where Does it Point?

If all this advanced cellular technology had been described to biologists a hundred years ago, they might easily have concluded that what was being described was futuristic alien technology out of a science fiction novel. The sheer quantity of specified information required for even a single protein of average length—say, three hundred amino acids—should give us pause.

Keep in mind, not just any amino acid sequence leads to a functional, folded protein. If we programmed the sequences at random—say by getting a parrot to peck away on a keyboard with twenty keys for the twenty amino acid letters—and then we checked to see what sort of protein the ribosome built from that random sequence, we would find that only the tiniest minority of resulting sequences would yield a folded and functional protein.

The number of possible combinations of three hundred amino acids is 10 to the 390th power—that is, 1 followed by 390 zeros. Of this cosmic-sized ocean of possible combinations, we're looking for those rare sequences that specify for the particular folded chain that yields a specific and indispensable protein machine, whose blueprint can be traced backward to the RNA template, which in turn has been copied from the DNA original. There do exist functional variants of this sequence, but the total number of functional variants is dwarfed by the vast ocean of non-functional sequences, much as some lines of functional software code might tolerate a few minor variations while remaining functional; but the total number of such options would be dwarfed by the universe of non-functional gibberish sequences possible in even just a few lines of text.

So just how rare are functional amino acid sequences in the larger ocean of possible sequences? According to researchers, studies have shown that functional amino acid sequences are as rare in 1 in 10^{24} to 1 in 10^{126} sequences,[12] with the functional sequences for an enzyme of typical complexity being about 1 in 10^{77} of the total possible sequences.[13]

To get a sense of just how rare such functional sequences are, imagine a planet Earth made entirely of sand, right down to the core. You're blindfolded and instructed to try to choose, at random and on the first try, the one specially colored grain of sand in the entire planetary sandpile, and then to repeat the same improbable feat on your first try on a similar sandy Mars. Your odds of pulling off that absurdly improbable feat are more than a billion times better than 1 chance in 10^{77}. The long odds against doing this give us some idea just how rare functional amino acid sequences are compared to non-functional ones. As with software programs, the rare amino acid sequences that work exist within a vastly greater sea of sequence possibilities that accomplish nothing useful.

But of course, cells are not composed of just one single protein. For a functional cell, multiply by many thousands the astronomically long odds against finding that single functional protein sequence, and we begin to grasp the challenge for unguided, unintelligent evolutionary processes to write all the high-tech code required for cellular and advanced life. And understand, most proteins are not like luxury options on a car; they're basic to all life functions. They play a myriad of roles in building vital cell structures and performing hundreds of chemical tasks necessary to the cell's survival.[14]

How do these many different kinds of proteins team up to do all the work that needs doing? Scientists have cataloged protein function, structure, and working relationships, and found that the average size of a team of proteins working together is six. Lone Ranger proteins are relatively rare.[15] All of these teams, in turn, are part of a symphony of hundreds of teams, each performing a specialized function. Taken together, we see that the appearance of design in molecular biology doesn't just reside at the level of a single precisely coded protein machine, but also at the group level, where various protein types are precisely matched and coordinated.

There are so many protein types and protein functions, and so many teams of proteins, that the study of them has given rise to a new branch of biology and a new object of study. The work of mapping out all the protein form and function in a cell led to the birth of the branch of biology called proteomics.[16] And the map showing the numerous protein interactions in various teams is called the interactome.[17]

Our key takeaway is this: The most striking thing about all three macromolecules—DNA, RNA, and proteins—is that they carry within their molecular structure something that goes beyond physics and chemistry, something we are quite familiar with in other areas and have touched on above: information. To understand why both philosophers and information theorists view information as something more than physics and chemistry, consider this: The paragraphs and chapters of this book can be stored on book paper, in a personal computer, or in the cloud. The physical medium can be wildly different, but the content of the book, the information, remains the same. The same goes for the information required to convey your favorite song. It could reside on a digital file in the cloud, in the memory on your phone or computer, or on a vinyl record. The information required to replay the song is stored in a physical medium, but it is not the physical medium. It's something more—something immaterial.

That by itself was a revelation. But what has shocked scientists in recent years is the discovery that genetic information is just a part of a much larger system of coded information spread across the cell in all its dynamic interactions. Scientists have stumbled upon a well-orchestrated ocean of cellular information beyond DNA, RNA, and proteins.

For decades, biology students were taught to memorize a little motto that summarizes what is known as the Central Dogma of molecular biology: *DNA makes RNA makes proteins*.[18] Many are left with the impression that this is the whole picture. Far from it. For instance, as we saw earlier, many segments of DNA produce RNAs that are never used as blueprints for proteins, but instead go to work inside the nucleus. In the wake of such discoveries, Stanford scientist Stephen Quake suggested that the "Central Dogma" needs to be replaced by the "Cellular Dogma,"[19] a new formulation that makes room for these revolutionary discoveries. The epigenome beckons. It's time to enter this new frontier.

3. A Voyage Through the Cell

As we saw in Chapter 1, the startling results coming from ENCODE in 2007 and 2012 revealed that 80 percent or more of the human genome is biochemically functional after all. Much or all of the DNA that does not code for proteins turns out not to be junk, as evolutionary theory had predicted. Instead, it is being assiduously copied in the nucleus and put to use. During this same period, biologists were waking up to another series of shocks—these from research on the informational realm situated above the genome, today known as the epigenome.

The once-dominant, gene-centric view is known as the genetic paradigm. The view replacing it regards DNA as akin to a filing cabinet full of information, with the epigenome performing the management functions that operationalize that filing cabinet of genetic information.

An early warning that the genetic paradigm was wanting came mid-century from American geneticist Tracy Sonneborn. As biologist and historian of science Jan Sapp explains in his Oxford University Press book *Beyond the Gene*, "Sonneborn became the most vigorous and skilled apologist for the biological significance of cytoplasmic inheritance and resisted what he considered to be the 'obsession' with the gene."[1] His experiments on what he called "cortical inheritance" caused a stir, but the ascendancy of the genetic paradigm meant that historical overviews of twentieth-century molecular biology tended to ignore or at least minimize such developments.

Figure 3.1. Stephen Quake, Lee Otterson Professor of Bioengineering at Stanford University.

As a corrective, in 2020 French biologist Paul Peixoto teamed up with three colleagues to survey several decades' worth of epigenetic discoveries, which they summarized in the *International Journal of Molecular Science* under the title "From 1957 to Nowadays: A Brief History of Epigenetics."[2] The discoveries reviewed there brought to light a shadowy system composed of layer upon layer of additional master control functions. This system does much of the directorial work for the genomic orchestra.

Bits and pieces of the story were known by the 1990s, but more recent discoveries have raised the curtain on—to shift the metaphor— a complex ballet of nanobot proteins writing and erasing the various epigenetic marks placed on DNA and on the tails of the histone spools where the DNA is wound. All these discoveries have pulled back the curtain, allowing us to see more of the dynamic molecular dance within the chromatin complex, where DNA is wound onto the histones.

Biophysicist Stephen Quake is a major pioneer in this area. He focuses on information flow in the cell, a promising avenue in the search for medically relevant cell therapies. Among many achievements, he perfected a method of "microfluidic automation that allows scientists to efficiently isolate individual cells and decipher their genetic code."[3] As part of the celebration of the fiftieth anniversary of the journal *Cell*, Quake was invited to share his assessment, a "look back and look ahead" at the field of cellular research. The resulting 2024 article, "The Cellular Dogma," has captured widespread attention.

In his analysis, Quake synthesizes a myriad of discoveries and explores their implications. "I will put forth what I see as a major conceptual challenge for biology in the next decade," he writes, "one that is inspired by Crick's Central Dogma: understanding information flow in the cell in the most general sense."[4] He then considers whether cell types could arise from genome sequence alone:

> Despite having sequenced hundreds of genomes of multicellular organisms, we still have no way to predict the cell types of an organism from the genome sequence alone. For clarity, by "cell type," I mean long-lived states of the cell and not the more difficult question of transient cell states, which can be subtle and change on short timescales. This is a very simple and fundamental question, and it seems likely that the genome sequence could contain all the information needed to answer it—but to date, it is not possible.[5]

Following this ambivalent summation, he takes a guarded step in the direction of open heresy against the Central Dogma:

> It is certainly possible that the hypothesis here is wrong, that the genome sequence alone does not define the cell types, and if this turns out to be the case, it would be a fascinating discovery. After all, the genome itself is a molecule with epigenetic chemical

modifications that aren't accounted for in the genome sequence, and the molecule is also packaged in chromatin [DNA wrapped around histone spools], which has its own chemical modifications and changes physical state selectively to make the genome accessible to transcriptional complexes and so forth. Perhaps there are proteins or metabolites that are inherited by the embryo from the sperm and egg, and those contain necessary information. Perhaps spatial distributions of molecules and cells encode necessary information to guide the trajectory of multicellular development.[6]

Then, after touching on various kinds of epigenetic information and what they might contribute, he makes a remarkable statement:

> I don't think there is a single example where naked DNA and a chemically defined in vitro solution has led to a living organism. As such, it would be very interesting to have a precise understanding of what and how much information is encoded in the genome sequence alone and what is stored by other chemical means in the cell, whether by epigenetic coding of chromatin, or via other biological molecules.[7]

This is momentous. Having summarized the old Central Dogma, he presents a new framework that brings in the totality of a living cell as the repository of coded information. That is, each sector or dimension of epigenetic information may be found to be crucial, with epigenetic information functioning as a network overseeing and selectively using the DNA files it needs.

So where in the cell is all this control information? What form does it take? What are its roles? And how precisely does it interact with DNA? To begin to answer these questions, let's descend into the bustling, watery world of a living cell.

20,000 Bots Under the Sea

Imagine you have been invited to a futuristic discovery center, a lavishly funded facility that has pioneered the ability to shrink people and objects many orders of magnitude. A wild fiction, to be sure. Even if such miniaturization were possible, any number of fine-tuned parameters of the laws and constants of physics and chemistry that make organic life possible would be thrown into havoc by the miniaturization. We would arrive on the other side decidedly dead. But that never stopped Ant-Man or the Magic School Bus or the Szalinski family in *Honey I Shrunk the Kids* or the crew of *Innerspace* or, for those readers with even longer memories, *The Incredible Shrinking Man* or the crew of *Fantastic Voyage*. So, let's climb into the incredible shrinking submarine with our tour guide and begin the journey.

The strange vessel is equipped with panoramic windows in every direction. As you buckle in, everyone is encouraged to close their eyes. A launch countdown sounds over the intercom. You feel a bit of vertigo as if a recklessly fast elevator were speeding downward from the top level of a skyscraper. There's a shrill metallic ringing and then both the sound and all sense of motion stops. A moment later you're jerked back into motion, but now the motion feels like that of a ship in roiling waters. Your piloting tour guide, seated with you and the other members of the tour in the spacious cockpit, raises a hand to signal that all is well. "They're moving us into position," he explains, "just a few billionths of an inch from the water-engulfed surface of a living cell."

The jerky motions of the ship give way to a steadier, more stately motion. "Okay," your tour guide says, "go ahead and open your eyes."

When you do, there before you, on the other side of the submarine's main viewing window, is an enormous cell—or at least it appears enormous, since you and the submarine have been shrunk down to a tiny fraction of the cell's size.

The vessel approaches the outer surface of the cell and slips into one of the many gates in the cell wall. Now all around you are the bobbing heads and trailing fronds that form the cell wall's double-layer lipid membrane. After squeezing through this dark curtain-like structure, your vessel passes into the cell interior, marked by a lattice-like maze of beams and girders. It looks as if some gargantuan and enormously intricate skyscraper were under construction beneath the sea—with all the interior pillars, trusses, and girders laid bare in a vast underwater cavern.

Your guide gestures with a sweep of the hand. "What you see all around you is the internal structural skeleton of the cell. Most of you are familiar with DNA, with *genetic* information. We'll encounter plenty of that. But some of the cell's most important *epigenetic* information is embedded here, in the precise way that these structural girders are positioned and attached to the cell membrane. This informational pattern is especially important in a newly fertilized egg—the zygote. Even when it is still just a single undivided cell, the zygote has far more information in its 3D structure than we imagined, far more than just what is rooted in its DNA."

After descending through a maze of tubular braces and beams, you spy ahead the nucleus, the spherical storehouse of the cell's DNA. At the tiny scale you and the sub have been shrunk to, it looms as a huge, shimmering globe, an underwater city whose surface teems with the traffic of complex molecules shuttling to and from its surface. As the sub draws closer, you see that the molecular blobs pass in and out of the nucleus via hundreds of large circular pores in the wall of the nucleus. Each portal has tiny hairs protruding around the entrance. The submarine approaches one of these portals and slips through.

Arrayed before you is a vast and crowded underwater arena. Hundreds of aisles penetrate the maze of stacked loops and clumps of DNA in every direction. Your tour guide pilots the sub into one of these narrow aisles and aims the sub's directional searchlight on a wound-up mass of DNA.

"You know those spools in sewing boxes, with the thread around them?" he says. "Here the DNA is the thread and the histone complex is the spool—only, as you can see, it's a lot less tidy than a sewing-box spool. Don't sell it short, though. It's an engineering wonder, enabling DNA to be compacted to an extraordinary degree. Each histone protein—reminiscent of a tiny clump of pasta—is a string of one hundred or so amino acids precisely folded into an irregular Z-shape. Once folded, eight of these histone proteins combine like a three-dimensional jigsaw puzzle to form the ideally shaped spool, called a histone core. In human cells millions of histone cores are produced as the cells divide and stand ready for the new DNA to be wound around them."

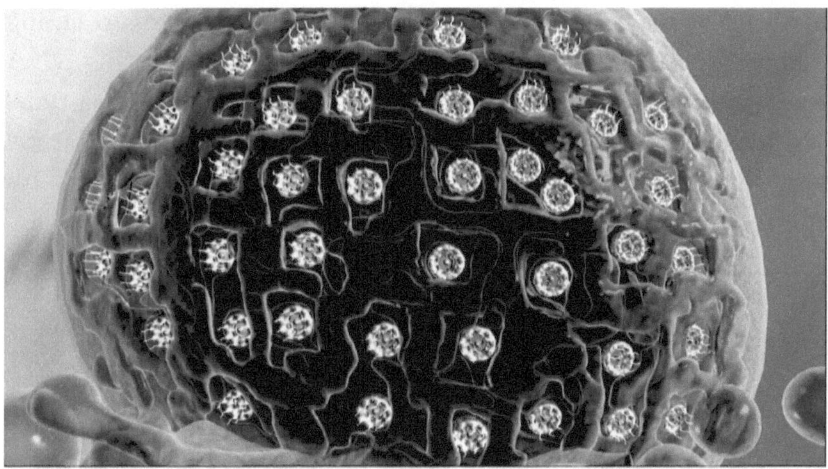

Figure 3.2. The nucleus, where in eukaryotes almost all DNA is stored. Note the gateways, called nuclear pore complexes, to enter and exit the nucleus.

Your pilot tour guide points the searchlight at something odd sticking out of a spool on one side. It looks vaguely like a thin branch, or a tail. As the sub draws closer, the pilot illuminates three other tails like the first one. "Each histone core is equipped with eight tails. You can spot four of them on the side facing us."

He narrows the spotlight beam and trains it on a strange Y-shaped tag attached to one of the tails that seems to be bulging out and pulling away from a histone bundle. "Note what is happening here," he says. "The little molecular Y-tag is causing the whole tail to be steadily pulled away from the layers of DNA on the spool. This is a key marker in the epigenome system. The tag is an acetyl molecule, attached to a precise spot on the projecting tail of the histone spool with the help of a special protein machine. This particular tag functions as an unlocking device. When the acetyl tags are placed on the tails, those tails swing away from the spool, and this makes it much easier for the file of DNA that is wound onto that histone to be pulled off and read. On the other hand, if the acetyl tags are removed by a different protein machine, the slender histone tails snug up closer to the DNA wrapped around the spool, allowing the tail to act like a clamp, securing the DNA tightly onto the spools and making it harder to be read. It's like a locking and unlocking system on a filing cabinet. These acetyl-tagged histone spools are all over this DNA warehouse. If we tried to count them all, we'd be here a long time. The information required to manage the precise placement of acetyl tags, something like a dynamic 3D map, is stored beyond DNA. It's epigenetic."

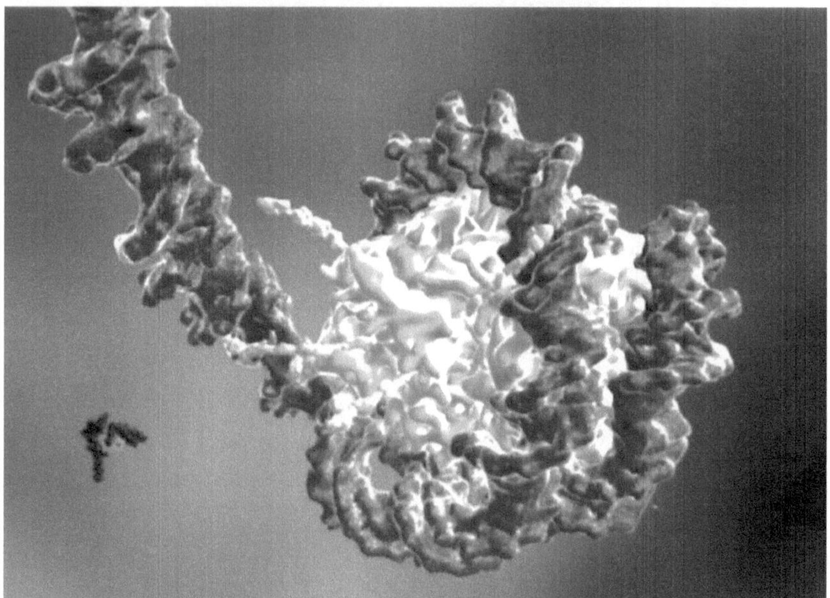

Figure 3.3. A histone core spool with two windings of DNA—called a nucleosome when the eight histone molecules of the histone core are wrapped with DNA. Histone tails (the white tails in the image) have many functions. On the left an HAC enzyme approaches carrying an acetyl tag, which will be attached to one of the tails.

Outside the submarine window, a protein machine suddenly swoops down and removes an acetyl tag from the spool. "That's a histone deacetylase enzyme, or HDAC, engineered to grab and remove tags," the pilot says. "Its complementary protein machine, a histone acetylase enzyme—HAC for short—attaches the acetyl tags. If all that is not complicated enough, there are four other chemical tags also found on those tails."

You sense a hum and a low vibration. The intensity builds steadily, and then there's a voice over the intercom. It's mission control. "Incoming. Take evasive maneuvers."

"Roger that," says your guide, and deftly maneuvers the submarine away from the DNA spool.

The ominous vibration persists, but then a broad smile breaks across your guide's face as he looks toward the far end of the narrow DNA corridor. "We're in luck," he says. "We happen to be in a section of Chromosome 17 where a key stretch of DNA is being opened and read. The vibration you're feeling is the rumble of the big protein machines lumbering along the stacks of DNA just ahead. They're unlocking and prying open very specific sections of the DNA spool clusters. We call these clusters nucleosomes. The goal of the machines is to gain access to the target cluster of genes."

The pilot seems to have forgotten all about the command to take evasive maneuvers. Instead, he aims the sub directly into the action, a frenzy of strange underwater robots scurrying around the tiny, interconnected islands of DNA. One slice of the action stands out. As the genetic corridor opens wider, several DNA-manager bots swarm down on the DNA, grabbing, twisting, and rapidly unwinding sections of it from their spooled-up position on the histones. Soon they have the DNA molecule unfurled.

At first, the double helix up close looks like fuzzy yarn, but as the sub draws closer, the DNA's elegant spiral-ladder geometry comes into focus. Its unique form looks like what is depicted in textbooks, except the rungs are knobby and thick, with almost no space between them.

The tour guide pilots the sub down along the spine of the DNA, and as they pass by, you get a close-up of the smooth surface of the DNA letter pairs, the rungs in the twisting ladder. Your guide aims the spotlight at one of the rungs. "Notice here and there, the smoothness of the rung pattern is marred by something sprouting from the rung—a tiny antenna-like appendage resembling a baby sprout of broccoli. No, it's not an unfortunate growth. The molecule is a chemical marker called a methyl tag. In its free state, as methane, it has five atoms—one carbon and four hydrogen. But the methyl group has one less hydrogen atom—three instead of four. So think of the methyl tag as a stripped-down methane molecule—a minor variation on the gas molecule burned by the trillions in any stove fired by natural gas.

"Researchers discovered this tag in 1948, and they learned that it gets attached only to the C letters in DNA's four-character alphabet, and only to some of them. Methyl tagging is one of the core features of the epigenome. Tens of millions of DNA rungs are tagged this way, and it was discovered that this tagging pattern is indispensable to cellular function. The specific letters getting tagged are chosen for functional reasons. Some molecular biologists describe the tagged Cs as the fifth letter of DNA. On this view, there is A, T, C, G, *and methylated-C.* That last letter is crucial. Typically, it acts as an off switch for the stretch of DNA where it's attached. If all those stretches were on all the time, we could expect a nasty multi-molecule pile-up. Some scientists estimate that as many as 200 million methyl tags have been placed on precise locations on the DNA in a single nucleus."

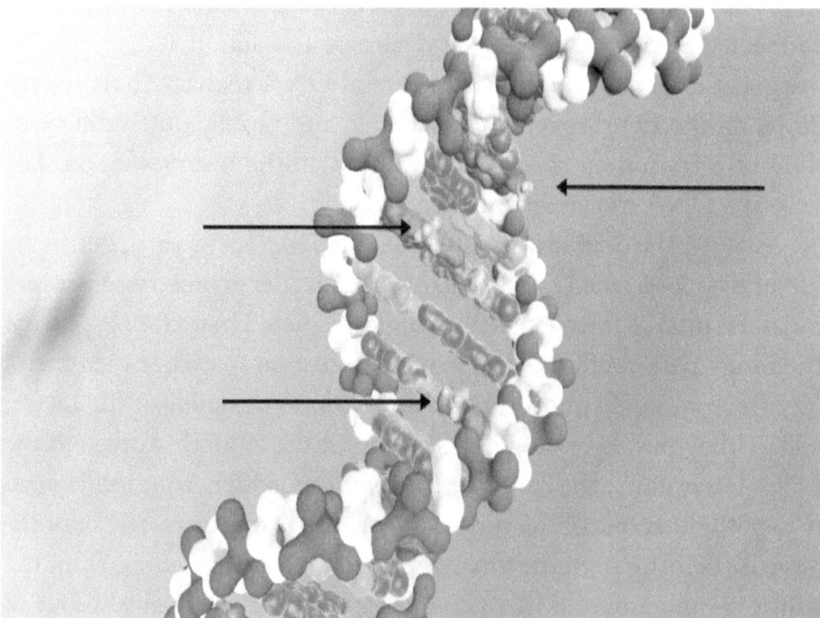

Figure 3.4. The DNA pictured here has methyl tags attached to certain C-letters. The methyl tags are the three small, attached molecules, each with three spherical ears.

The tour guide checks the sub's instrument panel and then resumes his commentary. "When a cell divides, the entire genome is opened up to be duplicated. All the billions of rungs of DNA are split open and replicated. So, do all these methyl tags get copied into the bargain?"

No one volunteers an answer beyond a reticent nod of the head from a fellow passenger.

"Exactly!" your guide replies, seizing on the weak gesture of assent. "The methyl tags are also copied over to the same corresponding rung in the new cell's DNA. The 'how' is interesting. The job falls to a special protein machine—a methyltransferase. But there's a catch. The pattern of methyl tagging varies from cell to cell, and while this methylation map is generally passed on unchanged from mother cell to daughter cell, there are untold thousands of exceptions, especially when an embryo is developing. The differences are essential, and the source of the differences isn't limited to DNA."

He checks the instrument panel again and then straightens in his chair. "Oxygen supplies getting a bit low. Time to make tracks."

If the lazy tune he begins whistling is any indication, he's expecting an uneventful return journey. But after your vessel passes through the lipid bilayer membrane, it is only past the cell wall for a moment before a second cell, an angry-looking bacterium, bears down on your ship.

Your pilot guide comes alive, but there's no panic in his movements. Just more of that childlike enthusiasm that has characterized his demeanor throughout the voyage. "Oh, handy!" he exclaims. "Oxygen tanks, low but not quite dangerously so. I think we can squeeze in one more spectacle before we surface. Let me maneuver alongside it here. An *E. coli* bacterium. These fellows live happily inside our intestines—by the billions—and do us a great service, helping us with our digestion. But it's not the bacterium itself I want to show you. Check out what's moving it."

You're close enough to the *E. coli* by now to see what looks like a loose braid of long hairs sprouting from its back, though unlike ordinary hair, these wiggle and twirl, functioning as propellers.

"Those whips are called flagella—amazing devices," the pilot says. "Machine-driven down at their base in the cell membrane. Essentially a sophisticated outboard rotary engine, complete with rotors, stators, bushings, a universal joint, and a drive shaft. Oh, and it can self-assemble. The motor has about forty parts, many with indispensable roles and each with a precise structure coded in digital form. Back of all that sophistication are some forty separate genes, packed with tens of thousands of base pairs of DNA. It's very sophisticated by the standards of human technology, but in some regards it's minor-league tech compared to what's going on inside the cell that we just exited."

A warning light begins blinking on his instrument panel. "Okay, now we really are low on oxygen," he says. "Not low enough to panic, but time to surface."

Why an Epigenome?

This imaginary voyage offers just a taste of the many actual wonders that researchers have coaxed from the cell. The difference is that instead of an afternoon tour in a handy miniaturized submarine, researchers have expended decades of painstaking labor unraveling the secrets of the cell via techniques such as X-ray crystallography and computer modeling. Let's review the factual content of our fictional submarine tour and fill in some details.

- During the descent toward the nucleus, we saw that a certain kind of epigenetic information was rooted in the exact placement of the structural girders and skeletal members of the cell, especially their precise points of attachment on the cell's membrane.

- We zoomed in on the methyl tags—tiny heads with three knobby ears. They are attached to some of the cytosine molecules, the C in DNA's four-letter alphabet, and the particular Cs are selected not randomly but in a very choosy way essential to function. The primary region where methyl tags are attached is a bit upstream from the start of the coding region of a gene, in the DNA pattern called the promoter sequence. This pervasive pattern of methyl tagging has been a prime focus of epigenetic studies since the 1990s.

- Early in the voyage we pointed out the dangling Y-shaped acetyl tags on some tails that protrude from the histone spool where the DNA is wound. These tags make it easier to open up the spooled DNA, enabling it to be read and copied. If the tags are removed, the DNA is wrapped more tightly on the spools, which sometimes is exactly what's required functionally. As was pointed out, specialized machines attach and remove the tags as needed, with the tagging system acting as a locking and unlocking system.

- During the voyage it was noted that there were four other kinds of tags attached to the histone spool, in addition to acetyl tags. One is the methyl tag, which we know about from the methylated C-letters of DNA. Interestingly, methyl tags are also attached at certain spots on the histone tails. The cell can attach two or even three methyl tags to the same spot. Those tail locations are said to be dimethylated or trimethylated. Each of these variations also has functional purposes—to either activate or repress gene expression, depending on the context with other epigenetic switches.

- The phosphate tags are a third kind of histone tag. This tiny molecule has a phosphorus atom in the center, connected to four oxygen atoms around its perimeter. Phosphates are key players in biochemistry, alternating with deoxyribose sugars (ribose sugars in RNA) to build the side sections of DNA's ladder and RNA's half-ladder.

- Ubiquitin proteins are a fourth kind of histone tag, dubbed that because they seem ubiquitous in cells. They are a well-known class of smaller proteins (a folded chain of seventy-six amino acids) that usually get attached to damaged or worn-out parts of a cell to signal that those parts are to be fed into the cell's recycling system. It was a fascinating discovery to find them used in a totally different way, connecting with the cell's chromatin structure to provide epigenetic control. There they connect with an amino acid called lysine on projecting tails of the DNA-wrapped histones. When attached to one key tail of the nucleosome, this tag encourages the silencing of the nearby gene, while this same U-tag, when attached to a different tail, encourages the gene to open up for transcription.

- The fifth kind of histone tag is the SUMO protein. SUMO stands for Small Ubiquitin-like Modifiers. But here "small" is misleading. More indicative is the acronym and its allusion to Sumo wrestlers, since SUMO proteins are somewhat larger than the similarly shaped ubiquitin proteins. To give some sense of scale, they weigh hundreds of times more than methyl, acetyl, and phosphate tags. These goose-shaped proteins act to block, or at least slow down, the transcription of genes wrapped around the tagged histones.

- In addition to the above team of tags, recent discoveries have brought other specialized chemical tags to light. See our endnote for a quick summary and an additional resource.[8]

We noted that acetyl tagging tends to open the spools. Interestingly, methyl tagging seems to close up or secure the spools. (We are simplifying things a bit here.) The addition of phosphate tags to the tails, and the linking of SUMO or ubiquitin proteins to the chromatin spools, also seems to affect whether the DNA is open for use or locked away securely. Sometimes, several tags work together in complicated ways on the same spool, like lines in a bar code conveying a complex set of instructions. In other words, according to current theory, the histone tags are not just locking and unlocking devices. They also work as signals, switches, and triggers that interact with the special protein machines scurrying around the packed spools of DNA.

A subset of these protein machines, called chromatin regulators, control the structure and function of chromatin. Proteins called transcription factors also play a role by recruiting either chromatin remodelers or histone-modifying enzymes.[9] These remodelers then open up selected stretches of DNA for copying into RNA when needed. This process is triggered when those factors bind to regulatory DNA sequences known as promoters and enhancers. A promoter is a regulatory region near a gene and influences when transcription begins. Enhancers are regulatory DNA sequences usually much farther away from a gene that, if bound to transcription factors, enhance the rate of transcription of a given gene. Sometimes we see multiple transcription factors interacting with the promoter section of DNA, which can bind even to enhancer signaling patches in the DNA that are quite far from the gene in question. The DNA-protein complex that assembles on the enhancer turbocharges transcription of a target gene. This complex, the enhanceosome, is a major discovery we will explore in Chapter 6.

Cancer's Ugly Shadow: The Enigma of Methylated Genes

One critical connection between epigenetics and human health has emerged—that between various cancers and abnormalities in the epigenome. Dr. Jean-Pierre Issa, formerly of the M. D. Anderson Cancer Center in Houston, notes that epigenetic switches controlling the expression of key genes can trigger cancerous conditions. For example, if a given gene helps modulate proper cell division, the cell depends on that gene to prevent the out-of-control cell division at the heart of all cancers. If histone tails "hug the DNA very tightly," Issa explains, "then it is hidden from view," and if permanently hidden, it can't be put to use. "It is the same as having a dead gene or a mutated gene. These are the kinds of things that can regulate gene expression and also become abnormal in cancer."[10]

We now know that changes in a gene, such as increased methylation (silencing) or reduced methylation (activation), can trigger certain cancers.[11] For example, oncogenes are normally benign genes that, when mutated or expressed at high levels, play a key role in turning a normal cell into a cancer cell, contributing to certain kinds of cancers.[12] In some cases, medical researchers have found that the normal silencing of such oncogenes is disturbed by faulty C-tagging of their DNA letters.[13]

Precisely how this happens is unclear, but some investigators think that some factor changes the pattern of methyl tags over time. A cell in healthy tissue is phased out through programmed cell death of older cells, called apoptosis. But the errant erasure of methyl-C tags in a defective cell allows the oncogene to "turn on and take off." Instead of dying at the proper time, the older cell is transformed by the oncogene into a fast-growing set of cancer cells. In other cells, methyl tags are unhelpfully added onto tumor-suppressor genes, silencing them and thus foiling their health-maintaining work of curbing tumors.

To recap, the epigenome's central function is to help control the expression of DNA, work involving thousands of genes in different cells. This role, when it goes awry, is connected to the eruption of cancers and other diseases.

Another role of the epigenome relates to the awe-inspiring process of growth and development of an embryo, beginning from the zygote. We turn now to this, the mother of all epigenetic wonders.

4. Mysteries of the Zygote Code

Ever since scientists first spotted the new molecular continent of the epigenome and began landing on its shores, strolling its sunny beaches, and penetrating its dark and mysterious forests, there has been no end of surprises. This voyage of discovery has been intimately connected with the question of how a stem cell—in our case, a fertilized human egg cell with its vast store of information encoded along the DNA molecule—can differentiate into all the types of tissue in a body.

For example, the human shoulder contains hundreds of billions of cells and various types of tissues, including muscle fibers, nerve networks, and bone structures. Every component is brilliantly aligned and interconnected to achieve exquisite functionality, with each tissue having its role to support and aid the joint's movement. The shoulder's formation begins modestly, at the molecular level. During development, each cell must perform flawlessly to ensure that the shoulder properly matures, even though the very first cell of a human being—the zygote that comes from the merging of egg and sperm—has a host of tasks radically different from building the bone, nerve, muscle, and skin cells of a shoulder. What enables the necessary cell differentiation and placement? What builds the complicated structural network? What sets up the unique life of each specialized cell within that system? As it turns out, the epigenome is key to this process.[1]

As development progresses, each cell is directed to unique use of its DNA files by the epigenetic system, which sits above the DNA while maintaining intimate contact with its genetic riches. This by itself represents a revolutionary discovery. But as scientists have made progress in pinning down the functions of the epigenome, another revelation has emerged. Each of us carries within his or her body more than two hundred versions of the epigenome. It has to do with the myriad of cell types in our bodies. Even on very conservative estimates, the human body contains hundreds of cell types (e.g., blood cells, skin cells, bone cells, nerve cells). And while they all have the same basic genome, each cell type has a unique set of epigenetic software, tailored precisely for that cell type. This fact may be the biggest shock to emerge from epigenetic studies.

Epigenetics researcher Berkley Gryder notes that if we include all the developmental stages of the cells in a human fetus, the number of separate epigenomes would be astronomical—in the many thousands. So, while the Human Genome Project had only one genomic informational system to map, a Human Epigenome Project would have a far more daunting task. If one includes in the count the differing epigenomes for the different transitional cell states, such a project would have to survey and map several thousand systems.

A daunting prospect, but scientists have responded to the challenge. As a start, the National Institutes of Health has overseen a project called the Epigenome Roadmap, whose opening phase mapped 111 cell types, while ENCODE scientists mapped another sixteen,[2] giving investigators a total of 127 epigenetic maps. ENCODE released its breakthrough findings in 2012 with the publication of about thirty papers in various scientific journals, including *Nature*, reporting that about 80 percent of the genome is biochemically functional.[3] Researchers investigated other cell types in the subsequent decade, cracking still other epigenetic codes.

Recent studies confirm that the control processes are directed from an information network outside DNA, directed in order to maintain health in the entire organism. The paradigm-shifting realization has now sunk in: DNA alone does not play the role of the cell's director. It is, itself, directed by a system higher in authority.

Revisiting the Construction Engineer

Though the terms *epigenome* and *epigenetics* were largely unknown to those outside the field of genetics at the start of the new millennium, by 2009 the epigenetics frontier merited an article in *The New York Times*. In "From One Genome, Many Types of Cells, But How?" science writer Nicholas Wade noted, "One of the enduring mysteries of biology is that a variety of specialized cells collaborate in building a body, yet all have an identical genome."[4]

As he went on to explain, scientists have concluded that in order to develop the brain, liver, bones, heart, and many other structures, there must be a different set of hereditary instructions above and beyond the DNA, acting to "open up" key lines of script in each cell, while making other lines functionally invisible.[5] Wade compared the varied tasks of human cells to a situation in which different actors read from the same master script while additional instructions, outside the script, block the actors from seeing the parts that do not pertain to them. Within a collection of living cells, the individual cells, analogous to the actors, do not see the entire DNA script—that is, the entire set of genes. The portions of the DNA script irrelevant to a particular cell are closed off by epigenetic markers.

As the geneticists quoted in Wade's article explain, the epigenome controls access to the genes, blocking many genes while allowing each cell type to activate its own special genes. Researchers have further concluded that the epigenome is involved not just in defining what genes are accessible to each type of cell, but also in directing the process of activating the accessible genes.[6]

Thus, the epigenome directs DNA to create many different types of tissues in the body from one stem cell. If each person has more than two hundred distinct versions of the epigenome and thus as many distinct orchestral conductors deployed in different kinds of cells, each with a distinct musical score, this raises crucial questions. How are the many different versions of the epigenome established in the first place? And how are they rewritten after the initial epigenome begins its directing activities in the zygote? What system directs the rewriting of the epigenomic system itself? To put it simply, how do so many epigenetic directors unfurl from a single, fertilized egg cell?

The Zygote Code

This is a complex process that must happen dozens of times, as cells are differentiated from their starting point. Based on responses from the geneticists Wade interviewed, he says the DNA rewrites the epigenome when needed, and the epigenome redirects the DNA usage in new ways as new cell types emerge.

Without completely rejecting this circular cause-and-effect pathway (we'll assume there's some truth to it), it seems an odd description of causality. What Wade does not mention is the foundational role of the zygote's interior architecture in producing the unique body plan of any higher species—whether a lily, a beetle, a kangaroo, or any of a million other life forms. The zygote's three-dimensional structure appears to be supremely important. Every molecule, every structural detail, every atomic nook and cranny of the zygote potentially contributes to the zygote's destiny. Some embryologists have now proposed that the precise molecular patterning of the interior of the zygote (in our case, the fertilized single-cell human embryo) is the locus of the supreme epigenetic code—the master code. Let's give it a name: the *zygote code.*

In exploring the informational mystery of the zygote for the present book, we have been guided in part by cell biologist Jonathan Wells, the late scientist who did pioneering work in this area. He earned a doctorate (his second) in cell and developmental biology at the University of California-Berkeley, specializing in embryology.

After publishing peer-reviewed embryology research while at Berkeley, he went on to write several articles and multiple books on evolutionary theory.[7] But what interests us here is his 2014 article in the journal *BIO-Complexity* challenging the widespread assumption that an animal's unique three-dimensional body plan could be specified from nothing more than a one-dimensional, letter-by-letter DNA code.

Figure 4.1. Embryologist Jonathan Wells.

When the zygote is poised on the threshold of cell division and embryological development, there is within that cell a structural goal—e.g., an elm tree, a lion, a hummingbird. Yet, where in the cell are the architectural plans for that goal embedded? Wells reviewed the sparse evidence other scientists had marshalled in arguing that DNA alone carries the plans for the spatial patterning of cells, plans said to lead to the construction of a new member of a particular species. He then described several fascinating embryo experiments that point in another direction, to certain structural patterns in the zygote carrying what he called "ontogenetic information."[8]

Inside the Egg Cell

To gain a vivid sense of some of these epigenetic discoveries, let's return to our fictional submarine for another miniaturizing descent, this time into an egg cell.

As the submarine descends, the only sound is the quiet hum of its propulsion system, but as the vessel gathers speed, a swishing sound mingles with the hum. Soon the egg cell looms ahead, looking like a small planet. This living planet looms larger and larger until it fills the entire view out the front of the submarine.

"We're heading for the famed zona pellucida surrounding the egg," explains your tour guide as he pilots the ship forward. "The ovum in some ways is far more complex than ordinary human cells. Notice that the surface, which appeared smooth a minute ago, is actually dotted with tiny tree-like shapes, with what look like thick hair braids hanging down on all sides. These chemical actors are called ZP3 glycoproteins.[9] A glycoprotein is a protein with sugars attached. Glycoproteins function as super-sensors and help orchestrate the fertilization process, allowing one and only one sperm to enter."

As you draw closer to the surface of the ovum, it first resembles desert sand, but when you are closer still, the consistency appears more like that of a bowl of oatmeal.

"You already know about the genetic code," the pilot says. "Another information-rich code is called the sugar code, carried in various complex sugars arranged just so. This sugar-based information system does much of its work outside the cell membrane. German scientist Hans-Joachim Gabius was a pioneer in the subject. He and his colleagues showed that these sugar complexes are extraordinarily diverse and provide 'high-density coding'—in their view, surpassing the information density of DNA and amino acids by a mile.[10] These sugar complexes often have branching treelike structures, but not for decoration. Those structures and their constituent branch molecules seem to carry their own language. Glycobiologists talk about this code having letters and words and built-in reading functions for cell-to-cell communication. The sugar code has a big impact on how proteins behave, on how cells react to stimuli. They're not just, well, sugar coating."

The submarine reaches the cell surface and begins pushing through the outer layer. You can hear the gentle swishing sensation of the thick lipid curtains surrounding the vessel.

After passing through the thick membrane, you find before you the vast interior of this living planet, held in place, like the vault of a cathedral, by long structural beams—scores of microtubules radiating out in all directions from the heart of the cell. The sub begins following one of the microtubules downward.

"These microtubules are the cell's superhighway system," the pilot continues. "Keep an eye out for the kinesins, tiny robotic deliverymen. Wait, there's one!"

The kinesin, on a microtubule ahead and to your left, is carrying a heavy load but strides along as if unencumbered. The long-legged thing looks like a droid out of a science fiction movie.

"Before we get too far from the cell surface, I want to pause to distinguish some things and show you one of the surprising keys to cellular inheritance," your guide says. "We passed through the zona pellucida, with its treelike ZP3 structures. Then came the cell membrane with its series of ion channels and other special proteins. And now I'm pointing our searchlight at the cell's cortex, with its forest of actin filaments and motor proteins. The cortex is the outermost layer of the cytoplasm. Dissolved nutrients and other chemicals in the cytopasm allow the cortex to function smoothly.

"The cellular cortex is a key part of the zygote code, another cellular information system. Patterns of vital information are embedded in the cortex, and it's now known that they can be passed on to daughter cells, even though the information is outside the cell's DNA. Tracy Sonneborn and his collaborator, Janine Beisson, were the first to prove this, with a trick they played on paramecium cells. As the cells neared the time to reproduce, Beisson and Sonneborn reversed the mounting orientation of the motorized oars protruding by the hundreds from the exterior of these cells. The DNA was completely unchanged in the process, yet those reversed cilia continued to face in their new direction in succeeding generations.[11] Researchers have coined the phrase 'cortical inheritance' to describe the phenomenon.[12]

"Two scientists who continued this research in the following decade, Stephen Ng and Joseph Frankel, described the cell as 'an architect' and said it 'not only makes use of the genomic information to produce the appropriate building blocks, but in addition, also arranges the building blocks according to the blueprints as defined in the pre-existing architecture."[13]

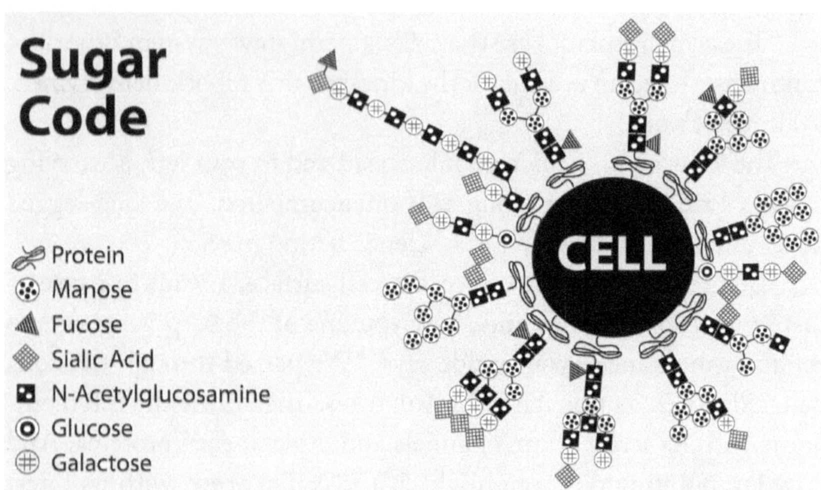

Figure 4.2. Glycans branching out like trees from a cell. Glycobiology studies patterns of glycans—sugar molecules—that perform key cellular functions beyond the cell membrane. This source of biological information has been called the "sugar code."

The miniaturized submarine continues downward, into the heart of the ovum. As you go, the microtubules draw closer and closer together, as if converging on a central hub.

"You can see that several dozen microtubules are delicately interlocking and interlacing here," your pilot comments. "Over to one side, you can see a large sphere with hundreds of round portals on its surface—the nucleus. It's one of the two round bodies at the heart of almost every kind of human cell. The other round body, much smaller by comparison, is the centrosome, the anchoring core for the microtubules. Inside its spherical body, two barrel-shaped structures are placed 90-degrees to each other, forming a T arrangement. We call these centrioles. Now, keep your eyes peeled and tell me when you spot the spherical centrosome."

You peer about in search of the centrosome. All you see is a complex lattice-like conglomeration, something like a spiderweb. A fellow passenger eventually breaks the awkward silence to comment that she doesn't see anything like the centrosome the pilot described. "Is it hidden on the other side of the nucleus?" she asks.

"No, it's not." The pilot smiles slyly. "I played a little trick on you. A human egg has a unique structure. It has plenty of microtubules, but it lacks a centrosome, which normally is the point of convergence for all those tubes. You may be thinking, isn't the centrosome necessary for the cell to divide and for a human zygote to grow into a fetus and then into a baby? Yes, indispensable. Cell division depends on it. To become a zygote, either the egg must develop a centrosome out of nowhere, or it has to receive one as a gift."

With that he swings the sub around to face the "ceiling" in the distance—that is, the inside of the cell's outer shell. It looks like the distant ceiling of a domed stadium, but is held up by the slender microtubules.

Now everyone's attention is drawn to one particular point on the ceiling, where a blister is forming. As the blister grows, the texture of the surface changes. Soon it develops into a big bulge, which grows until the bulge's central cap begins to melt away.

Figure 4.3. The spherical centrosome is an indispensable complement brought by the sperm, which quickly takes its full form after the turbine-like centrioles are moved to their proper location inside the sphere. Here we see reaching out in many directions the microtubules that are attached to the centrosome.

Where the cap was, a gaping hole appears, like the crater in the center of a volcano. Spilling through the hole is the unmistakable form of a centrosome—a small sphere with an array of stubby tubes jutting outward in every direction. Having penetrated the cell's outer shell, it begins to move toward the center of the cell, looking like a beautiful snowflake drifting downward. Following the centrosome through the hole in the bulge is a larger sphere, with tiny, telltale pores—the sperm's pronucleus.

Pilot and passengers watch in silence as the pronucleus and centrosome make their way to the center of the ovum and integrate into the architecture of the emerging zygote. "The sperm's pronucleus carries the sperm's DNA," says the tour guide. "What you're witnessing is the sperm's pronucleus merging with the ovum's pronucleus to form a fully operational nucleus. But without the sperm's other crucial gift, the centrosome, we would never see a true zygote form, or the baby after it. And this non-genetic contribution is just the showiest of a mind-blowing truth: *The 3D positions of billions of biological molecules in the zygote together contribute to the complete zygote code.*"

In this second voyage inside a human cell, we begin to glimpse the multilevel complexity found in the epigenome's storehouse of information. The living cell possesses vast riches of life-enabling codes, which go far beyond the spiral thread of DNA. Information, in a diversity of usable forms, operates in virtually every corner of the cell, from the outer cortex to the centrosome and its system of microtubules, to the histone spools with their decorated tails, to the methylation patterns attached to DNA. The mutual integration of these systems and their many layers of information is a marvel. Biologists will continue unraveling these complex relationships for decades.

5. Epigenetics and Health

Oncologists and geneticists are focusing much of their research on the connection between the epigenome and cancer.[1] This has led to an explosion of epigenetics-focused drugs. According to a major review article in 2024, many drugs recently approved by the FDA, or in trial stage, include pharmaceuticals that span almost all major categories of epigenetic health issues.[2] These attack the underlying causes of many cancers and other conditions—as, for example, a drug designed to treat a rare blood cancer, myelodysplastic syndrome (MDS).

Another momentous discovery in the field of epigenetics is the heritability of epigenetic changes. In Chapter 1 we introduced the work of Swedish geneticist Lars Bygren, who focused on family lineages in the sparsely populated agricultural region of Norrbotten in the extreme north of Sweden. His research revealed a link between patterns of binge-eating during bumper crop years and the introduction of a negative set of changes into the epigenetic systems of young boys during those same years. These epigenetic changes trimmed many years from the life expectancy of the children and grandchildren descended from the original group of overeaters.[3]

A similar line of evidence has been found in studies of mice in the laboratory of geneticist Randy Jirtle at Duke University. His work focused on the epigenetics involved in expressing a section of DNA called the agouti gene.[4] This gene causes a yellow coat to appear in mice. Unfortunately, this same gene can also be over-expressed, activating a biochemical cascade that results in obesity and diabetes. He found that one trigger was an exposure to high levels of a synthetic chemical pollutant in the environment, bisphenol A (BPA), used in some plastics and found in everything from dental sealants to water bottles. The chemical sometimes caused the epigenetic marks (methyl groups) to be removed from these genes, a phenomenon parallel to the unhealthy impact the binge eating had on the Swedish children in the Norbotten study.

An episode of *Nova* focused on Jirtle's research on agouti mice. After the program aired, Jirtle responded online to listeners, including those curious about the possible role of BPA in rising rates of obesity in humans. Jirtle said such a connection is possible but has not been proven: "There could be a connection between the increase in plastics in our environment and the rising incidence of obesity in humans. However, such an association will not be able to be demonstrated unequivocally until the expression and function of genes involved in human obesity are shown to be altered by BPA."[5]

Fortunately, that was not the end of the research story on the epigenetics of obesity in mice. Jirtle managed to produce healthy young mice in the next generation by changing the expression of their genes epigenetically—back to the normal, healthy setting. During gestation of the next generation of mice, he fed the mothers foods that provided high levels of methyl groups—mainly B vitamins, including folate and vitamin B12. As we saw in previous chapters, these methyl groups normally act like molecular switches, blocking the expression of certain genes. Jirtle's lab results were startling. The methyl groups wound their way, bit by bit, through the mothers' metabolic pathways, and were attached to the genes of the developing embryos. "It was a little eerie and a little scary to see how something as subtle as a nutritional change in the pregnant mother rat could have such a dramatic impact on the gene expression of the baby," Jirtle said. "The results showed how important epigenetic changes could be."[6] Or as he put it elsewhere, "The tip of the iceberg is genomics and single-nucleotide polymorphisms. The bottom of the iceberg is epigenetics."[7]

Figure 5.1. Pictured are a pair of mice: the yellowish, obese mother, and the darker offspring whose weight is normal. Epigenetic changes induced both health trajectories.

Four Basic Health Guidelines

Findings thus far are tentative, but biologists have gained a solid understanding of good habits that promote epigenetic health, a benefit that appears to extend to generations to come. Here are three do's and one don't for a healthy epigenome:

1. A *sound diet,* especially one rich in methyl-donor foods (see the table below) is important for maintaining a healthy epigenome.[8] In genetics classes at the University of Utah, students are now taught specific ways their diet can enhance their own epigenetic health and that of future generations.[9]

2. There is some evidence that a *positive mental attitude* can contribute to epigenetic health, especially as these attitudes reduce feelings of stress or enable one to handle stressful situations

once they arise. This finding parallels a large body of evidence linking improved stress management to general good health.[10]

3. Plenty of regular *physical exercise* and a proper level of body fat appear crucial for maintaining epigenetic health.[11]

4. *Smoking* has been shown by several studies to do extensive epigenetic damage.[12] The negative effects of smoking are among the most thoroughly documented in all of the epigenetic research literature. This agrees strongly with what we have already learned in the past fifty years about the effects of smoking on health.[13]

Methyl Donor Chemical	Food Source
Choline	Liver, Eggs, Fish (Salmon), Milk, Pork, Chicken, Broccoli, Navy Beans, Flax Seeds, Pistachios, Pumpkin Seeds, Sunflower Seeds
Betaine	Spinach, Broccoli, Cauliflower, Beets, Wheat Bran/Germ, Shrimp
Methionine	Turkey, Shrimp, Milk, Pork, Beef, Peas, Eggs, Brazil Nuts, Hard/Aged Cheese
Vitamin B3 (Niacin)	Salmon, Poultry, Beef, Peanuts, Peas, Avocados, Brown Rice
Vitamin B6 (Pyridoxine)	Liver, Tuna, Salmon, Chickpeas, Chicken, Potatoes, Bananas
Vitamin B9 (Folate)	Liver, Spinach, Broccoli, Asparagus, Chickpeas, Crab, Chicken
Vitamin B12 (Cobalamin)	Liver, Shellfish, Salmon/Trout, Beef, Yogurt, Swiss Cheese, Eggs

Figure 5.2. Some nutrients, and foods, helpful for maintaining a healthy epigenome.

Nurturing Behavior: A Factor in Stress?

The complex relationship between stress and epigenetics has become a hot area of research, including a Canadian study of stress in rats. Michael Meaney and Moshe Szyf of McGill University in Montreal have probed the relationships between rats and their offspring. Their study focused on mother rats and their grooming and nurturing of their young.[14]

Meaney and Szyf found that some mothers excelled at these behaviors, diligently grooming and nurturing their young; other mothers hardly attended to their young at all. Rats that had been groomed consistently as infants showed clear differences as adults. According to a report in *The Economist*, "Learning Without Learning," these well-groomed pups "grew up to be less fearful and better-adjusted adults than the offspring of the neglectful mothers." Also, "Crucially, these well-adjusted rats then gave their own babies the same type of care—in effect, transmitting the behavior from mother to daughter by inducing similar epigenetic changes."[15] That is, the behaviors were not merely learned. Nor were they driven by genetics changes. The driver, the researchers discovered, was epigenetic.

The second generation acted in similar nurturing ways toward their own offspring, producing the same epigenetic imprinting and epigenetically triggered behavior in the succeeding generation. Again, this shows that epigenetic changes, once initiated in a given generation, can be passed on to successive generations without changes in the genes themselves.[16]

One science writer concludes that Meaney and Szyf have made a "strong case that different epigenetic profiles resulting from early experience correlate with behavioral differences in adults," at least in rats.[17] What about in humans? Meaney and Szyf believe such a link might be demonstrated, and they have studied the role of epigenetics in the brains of those who commit suicide, compared to those who die in accidents. They are also looking for evidence of epigenetic differences in the cells of people who suffer from depression or have violent tendencies, research that could have major implications for the mental health field.

Another breakthrough came from Richard Hunter, a professor of psychology at the University of Massachusetts in Boston and a pioneer in the connection between stress and epigenetic changes. As a Rockefeller University report explained, he and his four co-authors found that "a single thirty-minute episode of acute stress causes a rapid chemical change in DNA packaging proteins called histones in the rat hippocampus... a brain region known to be especially susceptible to the effects of stress in both rodents and humans."[18]

When Hunter studied the specific changes in one key histone tail, he discovered something odd: The changes produced by the intense stress moved in opposite directions. Looking along the microscopic histone tail, he found that in one pinpointed location, the methyl tagging doubled, while in a different spot along the same tail, methylation dropped by 50 percent. From this, according to the report, the team concluded that "the sheer size of the change in histone methylation suggests that it is important to the brain's response to acute stress." The report then summarized Hunter's conclusions: "It's becoming increasingly evident, Hunter says, that the epigenetic changes like the methyl marks he observed and others, such as acetylation and phosphorylation, could play a significant role in the brain's response to stress and the treatment of stress related diseases, such as post-traumatic stress disorder."[19]

Another revealing study of the link between stress and epigenetics was published by Johns Hopkins researcher Richard Lee and colleagues. Their experiments, detailed in the journal *Endocrinology*,[20] involved feeding mice the hormone corticosterone in their drinking water. As explained in an article at *The Johns Hopkins Gazette* reporting on the study, corticosterone is the "mouse version of cortisol, a hormone produced by the human body during stressful situations."[21] This hormone altered the expression of certain genes, which in turn produced noticeably higher levels of anxiety during behavioral tests.

When the researchers probed one particular gene, *Fkbp5*, they found that there were "substantially fewer methyl groups attached to this gene" compared to the control group of mice whose drinking water had no hormone added.[22] So, through changed methylation levels, alterations were set in motion that in turn affected a "molecular complex that interacts with the glucocorticoid receptor,"[23] a key part of the normal functioning system of brain cells.

The Johns Hopkins research group found that the epigenetic tagging of the *Fkbp5* gene persisted for weeks after the mice were no longer receiving the hormone in their drinking water. This suggests the changes may be long-lasting. According to James Potash, director of the Johns Hopkins Mood Disorder Center, "This gets at the mechanism through which we think epigenetics is important." He continued, "If you think of the stress system as preparing you for fight or flight, you might imagine that these epigenetic changes might prepare you to fight harder or flee faster the next time you encounter something stressful."[24] Eventually, says Potash, doctors may have the technology and know-how to study human blood samples to detect the precise epigenetic changes connected to brain function. As a result, medical personnel might be able to design drugs that target epigenetic marks associated with psychiatric illnesses, providing another avenue of treatment, including for PTSD and some forms of depression.

Epigenetics and Exercise

In this chapter we reviewed just a sample of dynamic relationships between epigenetics and health. In addition, other surprises have jarred investigators, such as the relationship between epigenetics and aging. What appeared in the early 2000s as a dominant set of cause-effect linkages now appears twenty years later to be a much more complex and subtle network than was envisioned earlier. To trace this expanding circle of complexities would require an entire chapter.

Another surprise came as Romain Barrès, Juleen Zierath, and eight other scientists at the Karolinska Institute in Sweden published their findings on the epigenetic response of muscle cells to intense physical exercise. It has long been known that regular, vigorous exercise improves health and fitness. What their research uncovered was the role that rapid epigenetic changes play in this process. In their experiment, fourteen sedentary office workers were thrust into an intense hour-long bout on an exercise bike. Comparing thigh muscle biopsies taken before and after the session, a startling epigenetic change was spotted. Three genes involved in metabolism had been switched on, as methyl groups were removed on those genes' promoter regions. Ruth Williams reports in *Nature* on why this experiment was startling:

> The findings will come as a surprise to many researchers in the field. "Once a cell becomes an [adult] cell type, let's say a muscle cell or a fat cell, it is generally thought that DNA methylation is stable," says Ronald Evans, a molecular biologist at the Salk Institute for Biological Studies in La Jolla, California. "What Juleen is showing is that acute exercise changes the methylation status of the genome in actual muscle cells."[25]

Other researchers have confirmed this finding, showing that cells display a remarkably fast epigenetic response to bouts of exercise.[26]

6. Genomic Complexity: How Deep Does It Go?

A MAJOR FOCUS OF THIS BOOK IS THE EPIGENETICS REVOLUTION, but alongside that headline story is another momentous development. Researchers are finding that the human genome is itself bewilderingly more complex than anticipated. As one cell biologist put it in the journal *Nature*, the information processing in the cell is "infinitely more complex" than previously thought.[1] The article focuses on genome discoveries that, together, have revealed biology generally, and the genome specifically, to be "orders of magnitude" more sophisticated than previously believed. We scratched the surface of the genome's complexity in Chapter 2. Here we will launch out beyond that overview as we survey a series of startling new discoveries.

RNAs as Air-Traffic Controllers, Scaffolds, and Team Captains

As briefly covered in Chapter 1, ongoing research has uncovered a group of tiny genes that produce an array of super-small RNAs called microRNAs. These have on average about twenty-two half-rung bases, and don't code for proteins but do play several important roles. One is to attach to one end of a gene copy—an mRNA—to keep it from being translated into a protein when that would cause problems. This tagging action is like the grounding of an airline flight and thus has been compared to the action of air-traffic controllers.

The long non-coding RNAs called lncRNAs also don't encode proteins, but play similarly vital roles. They sometimes guide other DNA products to their proper destinations, again, much as an air-traffic controller in a tower guides the air traffic at a busy airport.[2] Just as the epigenome orchestrates the usage of DNA files with its various systems of tagging DNA and chromatin, so also many mysterious stretches of DNA, as well as once-overlooked RNA sequences, direct and regulate access to chromatin stashes of DNA and oversee the use of gene products throughout the cell's busy arena.

Scientists from Harvard and MIT's Broad Institute published a major study on lncRNAs, which painted a startling picture.[3] When interviewed about the study, one of its authors, John Rinn, formerly at Harvard University and Harvard Medical School, explained that the lncRNAs play an organizing role in cell differentiation from stem cells, "acting as a scaffold to assemble a diverse group of proteins into functional units." He added that lncRNAs "are like team captains, bringing together the right players to get a job done."[4]

A Revolution in the Idea of a Gene

Genes are now considered more than strings of DNA letters. They are clusters of diverse information modules, often though not always grouped in the same general location, but with each microcode having potentially a completely different purpose than an adjacent microcode. Think of finding a single super-page with twenty sentences, but then discovering that each sentence can be used, ignored, rearranged, or even reversed, depending on the needs of the moment. And sometimes sentences from other pages can modify the sentences on your super-page. This is but a crude analogy to how sophisticated and versatile a gene is.

Some scientists propose we can no longer maintain the older view that genes are discrete material objects, but that instead their essence is at the level of cybernetic ideas and information, and fundamentally not in physical matter substrates. Evidence of this emerging understanding can be seen, for instance, in a review article by Aaron David Goldman of Oberlin College and Laura Landweber of Columbia University. In the abstract they push back against the older view:

The genome is often described as the information repository of an organism. Whether millions or billions of letters of DNA, its transmission across generations confers the principal medium for inheritance of organismal traits. Several emerging areas of research demonstrate that this definition is an oversimplification. Here, we explore ways in which a deeper understanding of genomic diversity and cell physiology is challenging the concepts of physical permanence attached to the genome as well as its role as the sole information source for an organism.[5]

They go further in the conclusion. "These examples suggest a more expansive definition of the genome as an informational entity," they write, "often but not always manifest as DNA, encoding a broad set of functional possibilities that, together with other sources of information, produce and maintain the organism. Whether or not even this definition stands up to future discoveries remains to be seen."

Evolutionary biologist Richard Sternberg is more pointed: "DNA sequencing projects have revealed the gene to be a multilevel mediator of information that lacks a physical description."[6]

Files, Folders, and Overlapping Messages

Decades ago, geneticists were surprised to find that genes were arranged somewhat like files and folders on a computer hard drive. For example, many *E. coli* bacteria manufacture their own copies of a large amino acid called tryptophan and, to that end, are endowed with a handful of genes with instructions for building a team of enzymes that act in a precisely choreographed sequence to build up the cell's supply of tryptophan from simpler raw materials. These genes are arranged in sequential order in a folder called an operon, complete with a promoter sequence that is shared by this group of genes. Much of an *E. coli*'s genome is organized in such operon folders.[7]

That finding was eerie enough. Eerier still is the discovery that some genes contain overlapping messages in the same string of DNA letters. Imagine you've just read a long, complicated sentence in a book, and a friend stops you to point out that the author cleverly constructed the sentence and a bit of surrounding text so as to carry two quite distinct messages, depending on which letter you begin with and how you separate the string of letters into separate words—two wholly distinct sentences sharing most of the same text, and both sentences fully meaningful and functional in the context of the book. This is the sort of exotic function researchers are finding in DNA.[8]

MicroRNAs Meet Their Microprotein Cousins

We saw in Chapter 2 that DNA's hard drive contains various RNA genes essential to health. The microRNAs we just described as acting like air-traffic controllers are an example; but there is much more to this picture. In a *Discover* magazine piece, "The Sea Change That's Challenging Biology's Central Dogma," Gary Taubes explored the health role of microRNAs and summarized where the science seems to be headed. He then noted that researchers anticipate that "any disease that has a genetic component not yet identified—a long list that includes Alzheimer's, schizophrenia, bipolar disorder, obesity, heart disease, and diabetes—might be treated, at least in part, by adjustments in genes that code for microRNA." He added that when scientists ask "whether microRNA might play a role in a particular disease or health problem, these days the answer is almost invariably *yes*, because microRNAs appear to be everywhere, part of the underlying health of organs and crucial to biochemical cascades we only thought we understood."[9]

Also coming into focus in the 2000s and thereafter is the shadowy world of microproteins. These proteins are much shorter than one hundred amino acids and, because previously they went undetected, they have been dubbed the "dark proteome." Thousands of distinct microproteins have now been identified in the human genome, and some scientists say the number may reach tens of thousands before the survey is complete. It appears these microproteins, like microRNAs, are crucial to a healthy life.[10]

Matchmaking Across a Crowded Room

It also has emerged, as noted in Chapters 1 and 2, that genes can be edited to produce different protein outputs or (in the case of RNA genes) differing segments of RNA, all from one single gene. For example, from one fruit fly gene it is possible to make as many as 38,000 distinct protein products.[11] The outcome depends on which sections of that gene are kept in and which are snipped out.[12]

Part and parcel of such activity, some genomic research has found that a portion of DNA from one location on a chromosome may be linked to a snippet from an entirely different part of the genome (either elsewhere on the same chromosome or on a different chromosome) in order to produce the final result.[13] In some cases, we find evidence of different chromosomes bulging outward toward each other and seeming to have a genetic conversation, sharing segments of digital information to produce a new set of useful gene products. The level of integrated complexity is mind-boggling. It is further evidence of a highly sophisticated level of information storage and processing.

How is all this managed? It may require a long series of discoveries to get a handle on that, but one important clue has emerged, challenging what was once conventional wisdom. First, a couple of quick definitions. Exons are sequences of nucleotides (the As, Ts, Cs, and Gs discussed earlier) that are copied into messenger RNA for building proteins. Introns are intervening sequences of nucleotides that also get copied into messenger RNA but then get spliced out before the RNA is used to build a protein. Traditionally, exons were considered the information-bearing sections, with the introns viewed as mere spacers. But in recent years scientists have been surprised to find information embedded even inside the intronic spacer segments.

And the Gold Goes to… the Enhanceosome

Among the flood of new discoveries about DNA's complex control system, one marvel of engineering stands out: the enhanceosome. An enhanceosome, depending on the type, contains a half dozen to a dozen or more proteins. This protein-DNA complex switches on certain genes and, in doing so, acts something like a bicycle lock with numerous dials, only vastly more complicated. By far the most famous and well-studied of all these special structures is called the Interferon-Beta (INF-β) enhanceosome. It controls the copying of a key gene involved in our immune system.

An enhanceosome's function is remarkable in various ways. We can begin to appreciate this by comparing it to a unique type of protein, a transcription factor (TF), which attaches to a particular DNA sequence located outside of the coding region of a gene, but reasonably close—just upstream from its front door. This special sequence of DNA rungs is called the promoter. Once the TF protein is attached to that promoter sequence, it acts to recruit a copying machine (RNA polymerase) and associated factors to the transcription start site, where the process of transcription commences.

That's impressive, but the enhanceosome is even more so. Its role is similar to that of a TF protein, but instead of a single protein, it involves a whole team of proteins—typically around eight—scrunched together on a precise stretch of DNA. This team, after assembling on this DNA section called an enhancer, springs into action, recruiting still other proteins, called coactivators. This sets the stage for the copying machine to come alongside for its task.

If each team member binds to its designated "enhancer spot" in the DNA sequence, all goes well. But what would happen if, in all the hustle and bustle, one member of an enhanceosome team wasn't successfully maneuvered into its exact spot and fitted, like a three-dimensional puzzle piece, alongside the other proteins? The result is significant degradation or failure. All of them are required for proper function.

Daniel Panne, Tom Maniatis, and Stephen Harrison investigated the IFN-β enhancer and reported their results in the journal *Cell*. While the language of the article makes clear that the work was approached from squarely within mainstream evolutionary theory, the thrust of the article's findings was just how resistant the enhanceosome complex is to evolutionary change. "The nucleotide sequence of the IFN-β enhancer," they wrote, "is nearly invariant over roughly 100 million years of evolution, unlike the sequence of the gene.... Thus, the precise organization of the assembled transcription factors has had strong and continuing selective advantage. Moreover, mutational analyses have shown that virtually every nucleotide in the enhancer DNA sequence matters for some aspect of the response to viral infection.... The enhanceosome structure accounts for this conservation by showing that the transcription factors form a composite surface for recognition of the entire sequence and that adjacent transcription-factor-binding sites overlap."[14] Here then, we have the elucidation of a molecular system suggestive of irreducible complexity. This is a feature of some complex systems that, when present in living systems, poses a challenge to evolutionary gradualism, a matter we will explore further in Chapter 7.

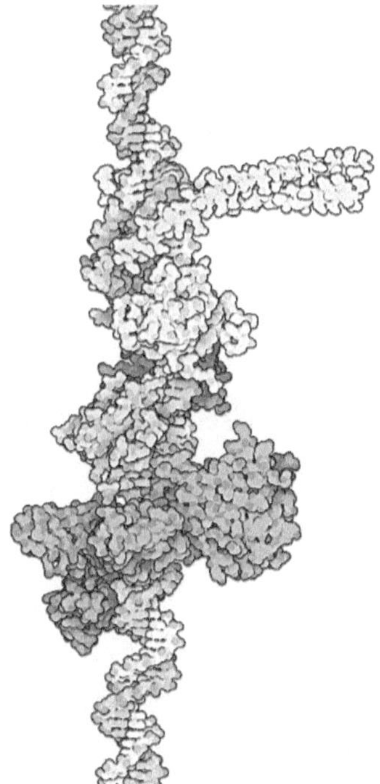

Figure 6.1. The IFN-β enhanceosome complex. The eight distinct proteins are each crafted to firmly grip part of the total DNA enhancer sequence.

Each enhanceosome team member possesses a unique design and a designated place in the larger puzzle of the enhanceosome where it alone fits. Their shapes are precise, enabling each one to latch onto specific DNA patterns. In other words, each protein in the enhanceosome team must be crafted with a hand-in-glove fit onto its designated part of the enhancer DNA sequence. Further increasing the challenge of all this is that the enhancer sequence tends to be located much farther upstream (or downstream) from the simpler promoter sequence, even thousands of nucleotide pairs away from the gene whose expression it controls. And yet it is able to maneuver into place and do its job in coordination with the faraway promoter.[15] The 3D folding of enhanceosomes is accomplished by multiple protein complexes that form precise DNA loops. The engineering sophistication evident here is staggering.

As we picture an enhanceosome complex hunched down across its enhancer stretch of DNA, one might wonder how rare or common these protein marvels are that coalesce on their designated enhancer zones. In humans, hundreds of thousands of enhancers have been identified across various cell types and tissues, with each individual cell type utilizing tens of thousands.[16]

The Rise of the C-nome

Before winding up our survey, there is one more layer of information inscribed in the genome that deserves mention, and this one was not glimpsed until recently. As biologists probed more deeply into how chromosomes are positioned, they discovered something startling. In 2023 eight Chinese scientists led by Hao Liu completed a massive study of the folded genomes of mammals and concluded that the genome's structural hierarchy appears to be precisely arranged, with the arrangement of the chromosomes relative to each other being anything but random. This has been dubbed the "C-nome"—with C standing for chromosome.

According to their findings, this highly specified hierarchical arrangement impacts generations of gene products and promises to shed light on many health disorders. As they explained, "Pathological mechanisms of diseases such as congenital developmental abnormalities and cancer… are attributed to alterations in 3D genome organization and aberrations in key structural proteins."[17] They also emphasized the importance of the 3D structure of the genome: "Linear DNA undergoes a series of compression and folding events, forming various three-dimensional (3D) structural units in mammalian cells, including chromosomal territory, compartment, topologically associating domain, and chromatin loop. These structures play crucial roles in regulating gene expression, cell differentiation, and disease progression."[18]

This recent wave of discoveries about the information infused in the 3D positioning of DNA began with a study of the yeast genome with its sixteen chromosomes. "Layered on top of information conveyed by DNA sequence and chromatin are higher order structures that encompass portions of chromosomes, entire chromosomes, and even whole genomes," the authors concluded in the journal *Nature*. "Interphase chromosomes are not positioned randomly within the nucleus, but instead adopt preferred conformations."[19] In *The Scientist*, Cristina Luiggi summarized the study's findings: "Understanding how DNA is folded, wound up, and packaged inside nuclei provides an additional layer of biological information to what's written in the base pair sequences."[20]

For the article, Luiggi interviewed National Cancer Institute biologist Tom Misteli, who hailed discoveries that led to a detailed 3D model of a yeast genome. "It's a fundamental property of the genome to be organized, to be folded in some way inside the nucleus," he commented. "Now it's becoming clear that there is more to the genome than the sequence. We have to describe how the genome is organized, figure out the mechanisms involved in the organization, and then figure out how the organization contributes to function. Tools are being developed to really address these questions in a systematic fashion."[21]

Oxford University's Eric Werner concurs:

> In the case of a building, the information for its construction and its structure lies not in the information that describes the parts that are used to construct it; rather it is in an architectural plan that is used by agents to construct the building. For humans... the information... resides in the genome, but not in the genes [alone]. Rather, it is in the network architecture that consists of coding and noncoding areas that determine the timing and spatial patterning of cells that ultimately results in the development of an organism.[22]

So another level of information has come into view within the genome itself: the architectural layout of the chromosomes and sections of chromosomes as they fit together in a grand, 3D jigsaw puzzle, albeit one that, once assembled, is more lively, creative, and capable than our most advanced robots.

Each genomic feature discussed in this chapter is an instance of a newly discovered bank of information. Taken separately, each would seem to constitute a line of evidence for a programmer, an intelligent architect of life—evidence that sits alongside the epigenetic data we have primarily focused on in these pages.

Taken together, this growing assemblage of evidence, genetic and epigenetic, would seem to signal not merely design but a strikingly ingenious design. For many who take all of this in, there is the strong impression of a superintellect at work. But is such an impression warranted? And is it in any way scientific? We turn to that question next.

7. Scrabbling for an Adequate Cause

L ET US RETURN TO WHERE THIS BOOK BEGAN, THE HUMAN GENOME Project. Launched in October 1990, the original director, James Watson, assembled an international team of scientists to tackle a complete recording of the human genome sequence. After a decade of assiduous work, a presidential press conference was held to announce a milestone in the history of genetics research.

As cameras captured the moment, Francis Collins, who had succeeded Watson as director in 1993, stood beside President Bill Clinton, with Prime Minister Tony Blair of Great Britain joining via a live television link. The project had been an ambitious effort, with $3 billion of funding from the US government, and with researchers forced to navigate many technical roadblocks along the way. But the project's network of scientists had overcome all these obstacles to successfully search out the precise letter-by-letter text of the human genome—the 3.1 billion letters of our DNA. A rough draft was being readied for publication, a roadmap of our genome showing the position of tens of thousands of known genes on the various human chromosomes.

One issue that was not addressed in the briefing: the origin of this vast library of digital information. Yet a few hints of God as the ultimate author were floated in the press briefing. President Clinton said, possibly with guidance from Collins, "Today we are learning the language in which God created life." Collins, for his part, said, "It is humbling and awe-inspiring to realize that we have caught the first glimpse of our own instruction book, previously known only to God."[1]

Was this just a bit of politically advantageous "God-talk," not to be taken seriously? In Collins's case, a theistic faith perspective was genuine. He had made no secret of his departure from the easygoing atheism he held to as a young scientist. Since coming to prominence as the head of the Human Genome Project in 1993, he did not hide from the press his personal story of reading the Oxford and Cambridge scholar C. S. Lewis to see what light he could shed on the possibility of God's existence. That was one of several factors that led to his abandoning atheism as a young scientist and embracing Christian theism.[2]

However, the moment wasn't quite an intelligent design lovefest. Collins, for his part, maintained near-unwavering allegiance to neo-Darwinism, holding that God appears to have fine-tuned the universe and possibly also the first single-celled life so that the neo-Darwinian process of purely random genetic mutations and natural selection could then do all the work of generating life's diversity, with perhaps only humanity's sense of a transcendent moral law requiring any additional direct input from God for its origin.[3] Collins also explicitly distanced himself from the intelligent design movement.[4]

James Watson's position was even less friendly to the design hypothesis. Three years after the news conference, he editorialized thus in his co-authored book *DNA: The Secret of Life*:

> The rift between tradition and secularism first opened by the Enlightenment has, in more or less its present form, dictated biology's place in society since the Victorian period. There are those who will continue to believe humans are creations of God, whose will we must serve, while others will continue to embrace the empirical evidence indicating that humans are the product of many millions of generations of evolutionary change…. It could be that as genetic knowledge grows in centuries to come, with ever more individuals coming to understand themselves as products of random throws of the genetic dice—chance mixtures of their parents' genes and a few equally accidental mutations—a new gnosis in fact

much more ancient than today's religions will come to be sanctified. Our DNA, the instruction book of human creation, may well come to rival religious scripture as the keeper of the truth.[5]

This was in the book's coda, which Watson concludes by scolding the 1997 Hollywood film *Gattaca* for asserting that "there is no gene for the human spirit." He calls this protest against genetic determinism and genetic reductionism "a dangerous blind spot."[6] The irony is that the book in which he makes these confident assertions, *DNA: The Secret of Life*, appeared at almost precisely the moment when the notion of DNA as the be-all and end-all "secret of life" was becoming passé, and as the wave of evidence against the "random throws of the genetic dice" scenario for biological origins was on the rise.

Figure 7.1. James Watson. Helped by insights from Rosalind Franklin and others, he and Francis Crick discovered DNA's double helix structure in 1953.

Inferences to Design

During the ten years between the launch of the Human Genome Project and the news conference celebrating its triumph, the methods of detecting the action of a conscious, intelligent agent or agents—long employed in everything from criminal detection to archeology to code-breaking—was further developed and formalized in a 1998 Cambridge University Press monograph, *The Design Inference: Eliminating Chance through Small Probabilities*. Hailed as a scholarly breakthrough by distinguished experts in statistical analysis, the book's methodology was further refined in a second edition (2023), in which the original author, William Dembski, was joined by software engineer and fellow information theorist Winston Ewert. As the book explains, one can detect the prior work of a designing intelligence (human or otherwise) by asking and answering two questions.

The first step in this two-stage design filter is to ask whether the event in question is extremely improbable, less probable than one chance in 10^{150}. Dembski arrived at this figure by calculating the probabilistic resources of the entire universe since the Big Bang, billions of years ago. The reason he focuses on probabilistic resources is that some extremely improbable events that seem impossibly unlikely become possible if there are enough tries. A newspaper might report that a lucky gambler in Las Vegas drew four of a kind in two consecutive hands and hint that there was cheating or poor shuffling of a new deck of cards. Though the probability of this event is about one in 17 million, there are millions of poker hands dealt in Las Vegas every year, maybe even hundreds of millions, so improbable events such as being dealt two consecutive hands of four of a kind not only can happen but can be expected to happen from time to time.

But what if a gambler was dealt, say, four royal flushes in a row? You can calculate the odds of that, find out how many poker hands had been dealt in the entire world since the invention of poker, and reliably conclude that cheating (a form of intelligent design) was involved. There simply are not enough probabilistic resources—enough tries—to tame this wild improbability.

So, is there a probability so slender that there are not enough probabilistic resources in the entire universe, stretching back to its beginning, to render the seemingly impossible outcome possible? This is Dembski's universal probability bound. In an exercise in hypergenerosity, he pictured all the elementary particles in the universe as something like independent slot machines randomly "seeking" the result in question. He further allowed that each of these particles could make attempts at the rate of what is known as the Planck time, the most minuscule theoretically measurable unit of time, about 5.39×10^{-44} seconds. Of course this method of calculating probabilistic resources vastly overestimates the probabilistic resources available for any real event. Dembski placed the threshold absurdly high in order to unequivocally eliminate the possibility of a false design inference, thereby allowing an investigator who is employing the methodology to be confident of any positive result.

The second question in Dembski's design filter concerns whether the event matches an independent functional or meaningful pattern. Or to use Dembski's language, is it *specified*?

A brief illustration. If one came upon a bunch of Scrabble letters strewn seemingly at random on the living room floor, the precise arrangement of these tiles would be extraordinarily improbable. After all, one could dump Scrabble letters, over and over again, until both death and taxes were abolished and not get that precise configuration a second time. But because the pell-mell arrangement of the letters does not match an independent functional or meaningful pattern, there is no reason to infer intelligent design. A dog or toddler may have simply knocked the Scrabble box onto the floor, scattering the tiles willy-nilly.

What if, however, you glanced back at the scattering of tiles and noticed that three of the tiles lined up to spell the word "CAT"? That's a specification. It fits an independent, meaningful pattern—namely a common English word. So you might informally infer that someone had purposefully arranged three of the spilled tiles so as to spell the word. If there was a child in the house who had recently learned to spell simple words and was enamored of cats, you might be even more inclined to such a conclusion. However, the design filter that Dembski formalized wouldn't trigger a design inference in this case. The odds of three of the tiles coming together to spell a common English word are nowhere near slim enough to reach the threshold of a mere one chance in 10^{150}—that is, one chance over the number 1 followed by 150 zeroes, a number that dwarfs large numbers such as a billion (nine zeroes) and a trillion (twelve zeroes). Here Dembski's filter may have delivered a false negative. Maybe the word "CAT" was purposely spelled out, but Dembski's filter would miss that. This is a limitation of Dembski's formalized method of design detection, but designedly so because, again, the emphasis isn't on avoiding false negatives—i.e., not recognizing that something was actually designed. It's on avoiding at all costs false positives—i.e., avoiding the mistake of concluding that something was designed when it really wasn't.

Now imagine that you left the house to run an errand and, upon your return, found that the Scrabble letters were arranged on the Scrabble board in a crisscross pattern of English words. No gibberish. Every tile precisely inside a square. Even without calculating the odds, you would know assuredly that this was the result of intelligent design. And if you did conservatively estimate the odds of such an arrangement, factoring in not only the odds that the letters would all spell English words but also that they would be precisely set on the Scrabble board according to the rules of the game, you would find that the odds were less than one chance in 10^{150}. Dembski's formalized method would thus trigger a design inference, corroborating the inference that anyone with common sense would have made intuitively.

This formalized method of design detection, as you might have guessed, is most useful in contested, mathematically tractable situations rather than in obvious cases of design such as the one immediately above. And in practical applications, investigators regularly opt for a less stringent probability bound, particularly where the probabilistic resources for achieving the outcome are obviously quite limited. Thus, imagine you returned home from your errand and found, instead of Scrabble letters spelling out "cat" or a completed Scrabble game, the following Scrabble sentence: TAKE OUT THE TRASH AND WALK THE DOG BEFORE PLAYING SCRABBLE. Seeing this, you would rightly infer that a housemate had, in playful fashion, left a message. That the message might not surpass Dembski's extraordinarily stringent universal probability bound wouldn't matter to you, for two reasons: First, in all of our experience with natural (chance or law-based) mechanisms, they never produce that kind of information—it always comes from a mind. Second, both you and Dembski intuitively know that the available probabilistic resources in this case are vastly smaller than those of the entire universe—so much smaller that you can immediately and confidently infer that this highly specified arrangement of Scrabble tiles is no chance arrangement but instead was intelligently (and cheekily) designed.

The Design Filter and Biology

The final chapter of *The Design Inference* investigates the connection between Dembski's formalized design detection method and evolutionary biology.[7] All forms of modern evolutionary theory appeal to some combination of chance and law-like processes, including natural selection. Some put more emphasis on one or the other, but all depend at some level on chance. Dembski's methodology provides investigators a tool for assessing whether a given formulation of evolutionary theory requires too much chance to be credible.

While Dembski was working on the original edition of his monograph, the work of formally detecting intelligent design in biological systems was also being carried forward by biologist Michael Behe. Behe's 1996 book *Darwin's Black Box* argued that intelligent design is indicated by the "purposeful arrangement of parts,"[8] seen in irreducibly complex molecular systems such as the bacterial flagellum and the blood-clotting cascade, the latter of which prevents us from bleeding to death from even small cuts. Such systems are like a mousetrap, Behe notes. Numerous indispensable parts have to be precisely shaped and fitted into place or the system fails. Remove even one of those parts and the system ceases to function.

We met the bacterial flagellum in our voyage in Chapter 3. This miniature rotary-propelled outboard motor, embedded in the membrane of *E. coli* and other bacteria, requires numerous essential parts, each precisely tailored for its job. The motor won't function (or is never built) if one of these is missing. So how did it evolve, tiny step by tiny evolutionary step, over many generations? Natural selection would never favor such a developmental pathway because natural selection lacks foresight, goals, and any patience for the non-functional. This challenge has become the focal point of a sometimes heated public debate.

Figure 7.2. Lehigh University biologist Michael Behe, author of a quartet of books that present the biochemical case for design, argues that irreducible complexity and a "purposeful arrangement of parts" are a hallmark of intelligent design.

Behe's concept of irreducible complexity poses a challenge not just to evolutionary scenarios for the bacterial flagellum, but to any evolutionary story dependent on blind, gradualistic processes, since natural selection, again, would quickly eliminate any dysfunctional intermediate step along the way to the functional system in question. At the same time, Behe argues, common experience shows us that intelligent agents readily form and arrange many disparate parts to build irreducibly complex devices. Given this combination of negative evidence against blind evolution and the positive evidence for the causal power of intelligence, Behe concludes that intelligent design is the better, more reasonable explanation for the origin of such biological systems.

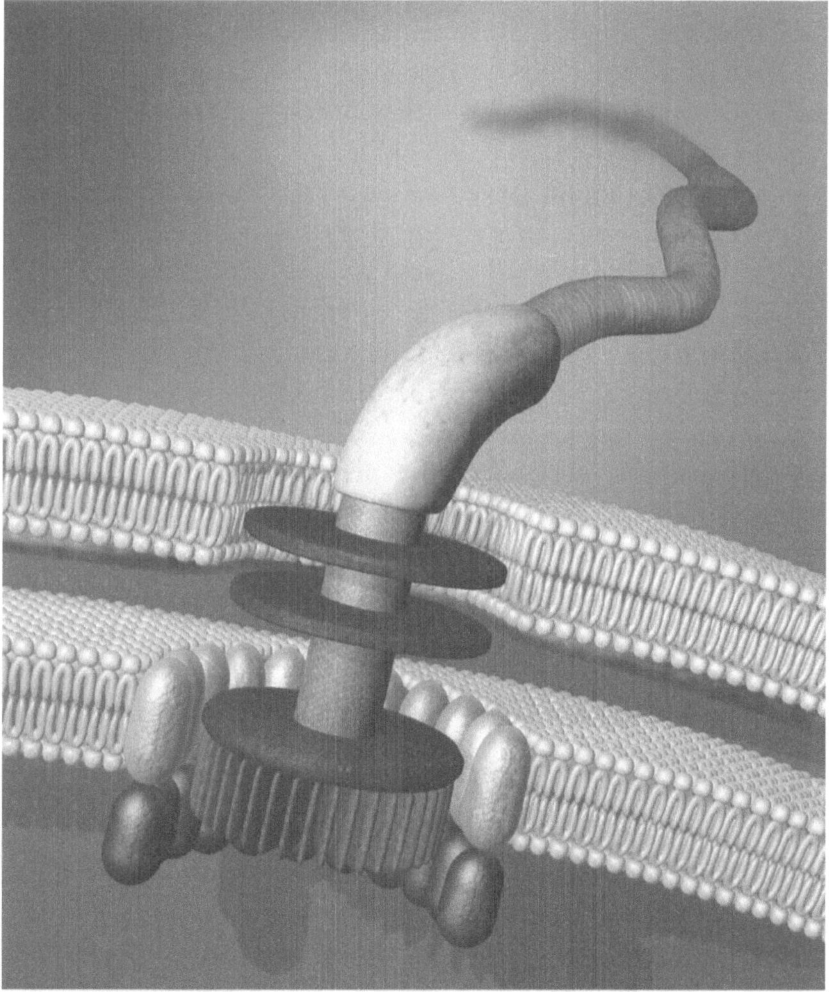

Figure 7.3. The bacterial flagellum is a marvel of molecular engineering, possessing numerous complex and distinct protein parts, with some twenty or more of them indispensable for its construction and function.

Some have attempted to rebut Behe's argument by appealing to the idea of co-option. This is the idea that the evolutionary process commandeers or combines simpler biological machines that were being used for different purposes. On this take, chance mutations are pictured as a tinkerer grabbing up whatever is at hand in the cellular realm on the way to evolving the completed system in question. Behe also addresses this idea. He shows that such proposals (including ones for the bacterial flagellum) are woefully short on both details and hard evidence. He further points out that, in any case, the work of co-opting and repurposing would require intelligent design in order to tailor and refit the co-opted parts, much as a skilled garage tinkerer must bring his know-how to bear in converting the parts of various machines into a new machine with a different purpose—for instance, converting a lawn mower and an old rowboat seat into a primitive go-cart.

Behe crafted another, if related, argument for intelligent design in *The Edge of Evolution* (2007).[9] There he investigated the limits of what random genetic mutations can achieve in various microbes, including the malaria parasite and antibiotic-resistant bacteria. He assembled a wealth of empirical data from the peer-reviewed biological literature and from this demonstrated that while chance mutations aided by natural selection can achieve some beneficial changes, the limits are quite severe, far below what is required to produce all of earth's rich diversity beginning from a first single-celled organism billions of years ago.

Behe followed this work with a third book, *Darwin Devolves* (2019), where he showed that a growing body of molecular biological data reveals that cherished instances of microevolution (e.g., polar bears evolving from brown bears) are cases of devolution—that is, genetic mutations breaking something at the cellular level in a way that creates an advantage for some offspring in certain environments.[10] The adaptive evolution is real, but because the changes arose from chance mutations breaking or blunting existing complex biological structures, the cases are a poor candidate for extrapolating to the evolutionary invention of complex new organs and systems. You can improve the gas mileage of a minivan by throwing out all the back seats, but you're never going to evolve the van by such methods into a boat or plane, or even a four-wheel-drive SUV. For that, you need intelligent engineering.

Published reactions to these three books were often negative, which opened the door for Behe to post or publish a stream of detailed replies. In 2020, he collected more than a hundred of those responses in *A Mousetrap for Darwin: Michael J. Behe Answers His Critics*.[11]

Philosopher of science Stephen Meyer also has produced a series of books developing the case for intelligent design. In *Signature in the Cell* (2009), he focused on the origin of life. In *The New York Times* bestseller *Darwin's Doubt* (2013), he explored the origin of novel body plans in the Cambrian explosion of new phyla some 530 million years ago. And in *Return of the God Hypothesis* (2020), he identified signs of intelligence in the fine-tuned universe as a whole.[12] More importantly, these three books develop Meyer's method of spotting design based upon an "inference to the best explanation." In short, when our experience shows that intelligent agents can and do produce the observed event in question, but that blind natural mechanisms cannot, we are justified in inferring design. Meyer's logic supports many of the examples of design inferences given above—such as our experience-based knowledge that a set of complex words on a Scrabble board must result from intelligence, or that only an intelligent engineer can retool machine parts and recombine them to suddenly produce an entirely new irreducibly complex machine.

The work by Dembski, Behe, and Meyer is joined by scores of other books and several hundred peer-reviewed articles that have added to the case for intelligent design since the White House announcement in 2000 celebrating the completion of the Human Genome Project. These writings have sparked a variety of responses from scientists and scholars. Many of the responses come from evolutionists, some of them outraged that anyone questioning modern evolutionary theory would be taken seriously. Some of these critics characterize ID as a threat to America's leadership in science.[13]

But in 2012 the eminent philosopher Thomas Nagel published *Mind and Cosmos* with Oxford University Press, and he placed the once verboten thought squarely in the subtitle: *Why the Neo-Darwinian Conception of Nature Is Almost Certainly False.* The atheist Nagel was not a figure who could be easily mischaracterized as an uninformed creationist crackpot. Then, in 2016, the oldest and arguably most prestigious scientific society in the world, the Royal Society, met to discuss a possible new path for evolutionary theory. And in the course of the meeting, various scientific luminaries frankly admitted that the neo-Darwinian synthesis was in a state of crisis.[14] After this, the rancor against ID subsided somewhat.

Even before this, Harvard University geneticist George Church was pushing for more open-mindedness in the evolution/design debate. Of Meyer's second book he commented, *"Darwin's Doubt* represents an opportunity for bridge-building, rather than dismissive polarization—bridges across cultural divides in great need of professional, respectful dialog—and bridges to span evolutionary gaps." Also notable was the reaction of paleontologist Mark McMenamin of Mt. Holyoke College: "It's hard for us paleontologists, steeped as we are in a tradition of Darwinian analysis, to admit that neo-Darwinian explanations for the Cambrian explosion have failed miserably. New data acquired in recent years, instead of solving Darwin's dilemma, have rather made it worse. Meyer describes the dimensions of the problem with clarity and precision. His book is a game changer."[15]

So, with all that as background, one may ask, where did the cell's genetic information come from in the first place? In answering the question, one might choose to foreclose any explanation that is out of step with philosophical materialism. Or one might choose to keep an open mind and consider two competing answers: (1) Nature's information arose gradually through the interplay of matter, energy, time, and chance alone; (2) Some sort of intelligent agency was crucially involved.

A key part of intelligent design methodology is to discern the uniform cause-and-effect patterns at work in the present and let these guide us in discerning the causes of various past events in the history of life and the universe. This uniformitarian principle needs to be carefully applied, of course. Here is how ID theorists use the uniformitarian principle in connection with DNA: In our uniform experience, we see coded language and information-packed strings of letters arising only from intelligent causes, never from mindless causes. It follows that coded language and information-packed strings of DNA arose from the same kind of cause—intelligence.

This conclusion follows the uniformitarian principle but also, and more broadly, a mode of scientific reasoning common to science, known as inference to the best explanation, elucidated by Peter Lipton at Cambridge University and discussed in an earlier chapter. Recall the Scrabble illustration in which you come home to the Scrabble sentence TAKE OUT THE TRASH AND WALK THE DOG BEFORE PLAYING SCRABBLE. This is just a string of forty-nine letters, ten spaces, and a period. Obviously it is an artifact of design, and yet its specified complexity is dwarfed by the highly organized genome of the simplest known bacterium (*Nanoarchaeum equitans*) with its genomic hard drive of 490,885 DNA letters organized with precision into 552 separate gene files.[16] And to progress from this bacterium (containing a half megabyte of genetic information) to one of the insect-like arthropods among the Cambrian fossils would require a jump to over a hundred megabytes of DNA information and thousands of genes. Are we to believe that all these new genes were produced by random throws of the genetic dice and preserved by natural selection?

One might appeal to the fact that Earth and the universe possess vastly greater probabilistic resources than Scrabble letters in somebody's home, but venturing there takes one onto thin ice. Douglas Axe's peer-reviewed work on proteins suggests that blind evolution lacks the probabilistic resources to generate even a single novel, functional protein type, never mind all the novel forms of life on our planet, each requiring scores of novel proteins working in concert.[17] And that work has been confirmed by the work of other researchers.[18]

Codes upon Codes: Meet the Supercomputer

This challenge to evolutionary theory is compounded by the discovery of the magnificently intricate epigenome. The epigenetics revolution means we now know that much of the biological information essential to cellular function isn't editable via random genetic mutations in DNA and thus is inaccessible to the Darwinian mechanism.

We find ourselves confronting numerous libraries of epigenetic information that work together to control the expression of DNA—libraries essential for life. This matrix of additional codes, grouped together under the epigenome umbrella, has come into focus only in the past thirty years. The sheer number of distinct biological codes that have been discovered is remarkable. Besides DNA's code, we see the methyl code, the histone code, the sugar code, the cortical inheritance and zygote codes, the gene body code, the origami code, and the enhancer code. The list of known biological codes, all of which must interact harmoniously for life to persist, has exploded. Two years ago, the German evolutionary biologist Robert Prinz noted this trend and produced a database of thousands of peer-reviewed articles from the past fifty years on biological codes.[19] The online resource lists 237 biological codes.

Is Prinz serious? Does he really mean that well over two hundred biological codes have been discovered? We should ask if all the codes he identified deserve to be called codes when fully understood. Prinz used a word-search function in exploring published articles, and certain biological patterns were called codes when the use of that word was meant perhaps only very loosely. Some of these biological codes are code-like in some respects but not true codes. Some function more like cogs in a grandfather clock's mechanism than like information-bearing letters in a code.

After making these and other adjustments to the list, we estimate that the number of distinct biological codes drops from 237 to about 160. Next, out of an abundance of caution, let's take that conservative estimate and cut that figure in half, giving us eighty codes. Now, for good measure, let's divide the figure by two again. That still leaves us with forty codes.

So we are confronted with at least forty interacting biological codes, coordinated symphonically to make life possible. This bespeaks a staggering degree of cellular sophistication. What cause can explain such a molecular universe of ultra-engineering other than a master architect of the system who created these interacting codes?

Berkley Gryder, working in an epigenetics research laboratory at a major university, describes the human cell as a "super computer."[20] He asks if mindless processes in nature have the ability to design and manufacture a supercomputer. To attempt to describe a cell's complexity, he sets it alongside a smartphone. The technology packed into these devices has progressed rapidly, so that smartphones now possess a functional density that even many science fiction novels set in the high-tech future failed to anticipate. But compare the functional density of this high-end cell phone to that of a human cell. If we represent a new iPhone's functional density as reaching to the top of a yardstick, the greater functional density of a human cell would reach some five hundred million yards high. The measuring stick would stretch into space—all the way to the moon and beyond. We submit that it is in no way irrational to infer that the best explanation for such technology is not a blind and mindless process but rather an extraordinarily intelligent architect.

DNA Tagging— Epigenetic Nitty-Gritty

Finally, there are several ways epigenetic tagging reveals the signature of a creative intelligence. First, consider the discovery of many different DNA tag maps (including histone tag maps), used to locate epigenetic information in individual types of cells. If the single genetic code arrayed in each human cell is impressive, what about the two hundred or more methylation code maps, a different one for each cell type? Recall that each methylation map in a typical cell in the human body has between one hundred million and two hundred million methyl tags, precisely placed on certain genes. And beyond those maps we have the challenge of mapping the methylation, acetylation, and phosphorylation of a cell's histones.

Recall also the irreducible-complexity challenge posed by biologist Michael Behe, considered now in the light of the cell's epigenetic tagging systems. Think about the methylation and demethylation processes, which switch genes on and off as needed, working with the millions of methyl tags attached to different sections of DNA in a human cell. If we study in detail the process that handles methylation, we find it brimming with examples of irreducible complexity.

One example is the complex methylation systems that place, remove, and transfer methyl tags on the C-letters of DNA. The process involves a team of distinctly tailored protein machines, such as the methyltransferase enzymes, working in concert to keep the methyl code in place as DNA is replicated. When a methyl tag is found attached to a C-base in a strand of DNA being replicated, the methyltransferase machine crosses to the newly constructed side of the double helix and settles onto the corresponding new C in that spot (within the CpG site—where a C is followed by a G on one side of the DNA strand). The machine quickly attaches a methyl tag to that adjacent C. In this way, copying of the entire genome is carried out at two levels: the genetic code (the base sequence of As, Ts, Cs, and Gs) and the methyl code (all methyl-tagged Cs).

What about removing methyl tags? An easy matter? Not at all. It involves several steps. When the need arises to activate a methylated (switched off) gene, all methyl tags attached to that gene must be removed. Here, a series of special proteins work in a tight sequence as they locate and hop onto the same methyl-tagged site in quick succession. The TET molecule, an enormous enzyme, starts the job. Three types of TET molecules are found in human cells, and they all have eighteen hundred or more amino acids each. When TET's three-stage operation is finished, it drops away, and a second protein called TDG swoops in to initiate the second phase of the removal. Once that is completed, one or more of the BER protein group comes alongside to complete the job.[21]

Now comes the irreducible complexity conundrum: How could this three-stage demethylation system work at all, if the necessary machines (specialized proteins) were not all there and ready to go as a coordinated team? All the essential steps require radically different protein workers to step up and accomplish their specific tasks, first to maintain the methyl code after DNA duplication, and then to place or remove tags when health conditions or cellular needs dictate. If there exists a plausible scenario for its gradual, piecemeal evolution, we would like to hear it. With a core of indispensable workers, it fully qualifies as an irreducibly complex system.

The same could be said of a host of other molecular machines that work together to manage epigenetic changes in other parts of the nucleus. Consider the dozens of delicate operations on the tails protruding from histone spools. Different teams of proteins are involved in tagging histone tails, entirely distinct from the ones we just saw. Let's consider how they manifest irreducible complexity.

Biologists now know that adding and removing acetyl tags requires a set of protein robots tailor-made for the job: the acetylases to add, and deacetylases to remove. What about adding or removing phosphate tags from the tails? A still different fleet of protein teams handles those jobs. Kinases place the phosphate tags, and phosphatases remove them. Finally, placement of methyl tags on the various histone tails is handled by another pair of protein teams tailored for the job—HMT and PRMT. Individual proteins in these teams can range in size from 700 to more than 2,000 amino acids. When it's time to erase methyls from a histone tail, removal is handled by other teams of large proteins, with chains that range from 850 to more than 1,500 amino acids. One of the most prominent teams is known as the KDM5 enzymes.

We have now seen seven distinct histone robot teams, each possessing its own set of proteins with their pre-programmed functions of writing, transferring, or erasing epigenetic marks. Each team is vital; each contributes to epigenetic health. These proteins all work together as an irreducibly complex team, where each protein plays its part. All are needed for the system to work.

And we should not imagine these large protein machines as simple blobs of disorganized organic material, akin to a congealed lump of spaghetti. Each consists of a 3D coil of hundreds of amino acids arranged in a precise sequence, with each protein type functioning like a very long meaningful string of characters in a book or software program. Each robot is assigned its unique task of adding, deleting, or transferring a specific kind of epigenetic tag.

Two facts confront us here: (1) Because of the significant length of each of these specialized proteins, they possess an enormous amount of functional information, very unlikely to be hit upon by dumb luck, with or without natural selection. (2) As noted in Chapter 2, the occurrence of functional sequences among the ocean of nonfunctional sequences for particular proteins has been explored experimentally and found to be extremely rare.[22] Axe, for instance, found it to be one in 10^{77} (a 1 followed by 77 zeroes) for a protein 150 amino acids long.[23] While that small probability is daunting, the odds of hitting on a much longer functional protein chain (say, two thousand amino acids) can be expected to shrink considerably further, to an unimaginably small and improbable number. Thus, an origin story for these proteins that is nature-only and mutation-fueled is implausible in the extreme.

Consider the limited time available for a random-search mechanism—say, 3.95 billion years as an estimated earliest appearance of life. That much time is but the tiniest fraction of the total time required for a blind Darwinian process to craft a single epigenetic robot, let alone the entire fleet of coordinated teams of protein bots that we have surveyed.

What's more, recent research suggests that these basic toolkits for epigenetic manipulation (writing, transferring, erasing) were already present in the original hypothetical ancestor of all creatures whose cells possess nuclei—that is, the last common ancestor of eukaryotes. A survey published in 2020 states:

> Eukaryotic organisms evolved complex epigenetic processes to orchestrate gene expression and genome dynamics. By applying a taxon-rich phylogenomic approach, including adding transcriptome data from several lineages of understudied microeukaryotes, we identify homologs of the epigenetic gene toolkit in diverse lineages across the eukaryotic tree of life. We show that gene families involved in chromatin modification and the processing of non-protein-coding RNAs originated in the last eukaryotic common ancestor.[24]

This startling conclusion puts a squeeze on the time available for Darwin's meandering process of undirected evolution to craft these fleets of epigenetic robots.

Taken together, the above findings are far from friendly to modern evolutionary theory. The rapid appearance of all this information-rich engineering sophistication points instead to the work of a supreme architect.

8. Peering into the Known and Unknown

The era of the epigenome has arrived. A vast new frontier lies before us. Scientists in many fields are working together to penetrate its secrets—to understand the size, shape, and complex interconnectedness of this vast information system. At the same time, we have seen how the DNA library is viewed with greater awe today than when the Human Genome Project was nearing completion a generation ago.

In this concluding chapter, we will reflect on what we have learned regarding the dance of genome and epigenome, and look ahead to the prospect of future breakthroughs in epigenetics. Finally, we will consider the wider implications of the extraordinary physical evidence explored in the preceding pages.

What Have We Learned?

In laying the groundwork for the book's central chapters, we explored the principal parts and systems of the genome, including its DNA, its RNA copies and, in human cells, the genome's roughly twenty thousand protein-coding genes. We also surveyed complex machines that involve the orchestration of multiple proteins into one system, such as the kinesin robot striding down a microtubule delivering a package, or that nano-engine, the bacterial flagellum.

We then delved further into the cell's sophisticated information system. Four aspects concerned localized informational systems:

methylated DNA, tagged histones, the centrosome with its associated microtubules, and cortex-based information, including the astonishing sugar code. The fifth aspect is not localized in any one part of the cell; rather, it encompasses the spatial positioning of all molecules in the architecture of the fertilized ovum known as the zygote. Various lines of evidence now suggest that the precise three-dimensional positioning of every biomolecule in a zygote contributes vital information to the myriad of developmental pathways of tissues and organs, leading to a creature's birth and eventual maturity.

We also looked at how epigenetics, every bit as much as genetics, is tied to our health. Changes to the epigenome are linked to aging and many diseases, including cancer, as well as to some mental health problems. Such epigenetic changes can be triggered by a diversity of causes, including stress, diet, chemical pollutants in the environment, and smoking. Ultimately, a wide variety of life habits and mental attitudes may prove to be linked with epigenetic health. This knowledge promises to bear positive fruit in the lives of future generations.

Finally, we touched on major developments in the understanding of biological information. Chief among these is the splicing code, the role of the 3D architecture of the chromosomes as a separate code, and the emergence of a new concept of the gene as something that transcends the physical, four-dimensional cosmos. What emerges is a transformed vision of the genome as not only fundamentally informational but also, at least in part, immaterial.[1]

A Look Ahead: What's Coming?

Many who have conducted research in epigenetics have confessed that, despite all the progress made since the early 2000s, we still face large areas of ignorance in this field. Some have described our current knowledge of epigenetics as equivalent to the state of science's knowledge of DNA shortly after Watson and Crick's breakthrough of 1953. If this picture is correct, we can think of the field as just now leaving infancy and ready to enter elementary school—a humbling thought.

One largely unanswered question concerns the extent of cross-linkage among the various informational layers in this dual system of genome and epigenome. Surely, many surprises await us, since surprises seem to be the order of the day in this field. As an example, in a study published in a 2020 issue of *Genome Research*, investigators reported stumbling onto a remarkable phenomenon. Placement of methyl tags on prostate cancer cells sometimes caused an up-regulation or switching-on effect.[2] The idea of methyl tags promoting a gene's activity runs counter to much of what the field had learned about methylation in recent decades, and it shows that epigenetics often depends on the biochemical context. This means that when we learn the rules of function, we also need to look for exceptions, since there are certain contexts—only vaguely understood at this point—that may reverse the normal effect of methyl tagging.

Also discovered was the presence of millions of stem cell methyl tags with unusual tagging patterns that affected the DNA in an odd manner. *ScienceDaily* reported that two findings stood out: In some cases, "DNA methylation appeared to enhance, rather than repress, the activity of the surrounding DNA," and there appeared to be "a role for DNA methylation in the regulation of mRNA splicing."[3]

Meanwhile, a series of research findings has revealed a fascinating new world of RNA methylation. Various roles were reported involving functional enhancement and control when RNAs have methyls attached; but generally this takes place on their "A" letters, not the "Cs" as with DNA. RNA methylation research continues to boom, work that promises to benefit the fields of medicine and wellness.

More fascinating still are hints of a link between the methylation code and the splicing code. This was borne out by studies in which brain-cell DNA found in worker bees and queen bees is radically changed by methylation. More than five hundred genes in those brain cells have differing methyl patterns, and the evidence points to the methylation as not simply silencing genes, but rather signaling to the cellular splicing system to cobble together, or re-edit, the pieces of genes differently, depending on whether the bee is a worker or a queen. The result of the studies was decisive: The embedded epigenetic code orchestrated the splicing process on genes, which yielded fresh protein results, depending upon the brain's changing needs. Since the two kinds of bees—worker and queen—require different behavioral patterns, it seems that hundreds of gene products in brain cells are snipped and spliced differently, according to two distinct master plans, so as to prepare the bodies and brain cells for the two radically different ways of life in the beehive.[4]

With such oddities and cross-linkages only now coming to light, we need to be aware of how very limited our knowledge of the dual genome/epigenome system is. We build on what we know, and as yet we have done little more than lay the foundation.

A Look Up: What's the Ultimate Concern?

As we delve ever further into the molecular biological realm, what do we continue to encounter around each successive corner? Specified complexity, irreducible complexity, and magnificent orchestration. We find ever more evidence of an architectural masterpiece. Beyond DNA we encounter the interacting control codes: the methyl code, the histone code, the zygote code, the sugar code, and the 3D code, among many, many others. A mechanism of random emergence of traits, selected by environmental pressures, may account for small-scale variations like those of polar bears, cichlid fish, and others described in Michael Behe's *Darwin Devolves*.[5] These unique forms arise when genes are broken or blunted in a way that happens to create an advantage in some particular environment. Such mutations, while fascinating, are inadequate to account for life's breathtaking diversity and sophistication. Only prodigious intelligence would seem to possess the palette, the paints, and the technical mastery to generate biology's many masterpieces.

By our lights, knowledge of a new epigenetic dimension of the cell's complexity adds immeasurably to one's appreciation for the beauty and ingenuity of life's design. Chance and natural selection fail to explain the marvel that unfolds when the epigenome takes a human zygote with its genetic stash and builds all the diverse types of tissues and organs needed to form a human body. As we consider its control functions that direct the use of different sections of DNA and precisely integrate all the cells, we find ourselves in the presence of a system so advanced that it bespeaks a seemingly limitless intelligence for its architect. And if we take account of the great variety of distinct life forms earth has ever harbored, each programmed with its own genomic and epigenomic codes, we can multiply this impression by the thousands and even millions.[6]

The orchestration of genome and epigenome, of all the diversity of biomolecules and cells, creates the beautiful symphony we call life. Richard Dawkins describes biology as "the study of complicated things that give the appearance of having been designed for a purpose."[7] Of course, as an atheist he sees this as an illusion and argues as much. In his view, natural selection is the grand illusionist, creating the mere appearance of design in biological forms. Yet natural selection has no foresight or long-term goals, as Dawkins himself emphasizes. Natural selection can only act on the modest digital changes presented to it, mostly in the form of point mutations in single letters of DNA. And as we have argued, there are sound, empirical, and mathematically tractable reasons to doubt that blind evolutionary processes, however conceived, can knit together even a few novel genes of DNA, or for that matter, even a handful of novel proteins. Given this, it is hardly unreasonable to doubt that any blind evolutionary process could code an entire genome and repeat the trick tens of thousands of times in the history of life. And to jump to a still higher level, why should we expect it to be able to create the multilevel epigenome in which we see the smooth interaction of forty or more codes?

Dawkins's *The Blind Watchmaker*, first published in 1986, frankly concedes that the details of organismal and cellular complexity not only appear designed but, more than this, "overwhelmingly impress us with the appearance of design as if by a master watchmaker."[8] If cellular complexity seemed to shout masterful design some forty years ago, before the epigenetic revolution, what about today, when many new features of our cells' information systems are found to transcend DNA while being beautifully integrated with it? When such systems are detected in every part of the cell, surely at the very least this strong impression of purposive design is only strengthened.

In the Oscar-winning 2001 film *A Beautiful Mind*, Russell Crowe portrays a Princeton mathematical genius, Nobel Laureate John Nash, who was blessed with a "beautiful mind" but was tragically plagued with schizophrenia. By contrast, the mind we see revealed in the genome and epigenome seems to us a beautiful mind at a vastly higher level—a mind that is, we suggest, the source of our own capacity for creativity and reason.

Or consider another illustration. If you ever tour the city of Barcelona in Spain, one stop will be unforgettable, one of the world's astounding architectural masterpieces. We refer to the Sagrada Família church. Its construction began in 1882, and now, after 144 years of meticulous crafting, it is nearing completion. Its mammoth four-armed cross was erected atop the main tower in February 2026, making it, at 566 feet, the tallest church in the world.

But its height is a mere footnote compared to the unusual design itself. In some ways it seems high-tech, as if equipped with a phalanx of huge rockets ready for launch; yet overall the design is earthy and organic, as if blending elements from the ceilings of Carlsbad Caverns with sculpted shapes resembling a dense grove in the Amazon jungle.

Every masterpiece can be traced to its master designer—the architect. Sagrada Família is no exception. It owes its elements of genius to one of the world's most talked-about architects of the modern era, Antoni Gaudí. His quirky, whimsical Park Güell and his undulating facades adorning his apartment buildings (like his La Pedrera) are legendary. Whenever you see one of Gaudí's creations, his exceeding genius and creativity radiate not only in the overall structure, but in the fine details that fill every nook and cranny.

In the realm of biology we also find dazzling architectural masterpieces, equipped with multi-level design details that exceed anything Gaudí attempted. These masterpieces also point to a master architect.

This inference to design is rooted firmly in scientific evidence and proven modes of scientific reasoning. At the same time, the inference, once made, invites questions that extend beyond science: If the universe was purposefully made, what was the point? If we are the product of purposeful design, what is the purpose of human life? Why are we equipped to ponder and pursue such questions?

Figure 8.1. The Sagrada Família church in Barcelona, Spain.

Recalling humanity to such questions may prove to be science's ultimate gift.

The fact that human intelligence, through scientific advances, can observe and decipher such intricate workings, and ask such profound questions, raises another question: Whence did that capacity for such intellectual pursuits originate? One explanation is that the human mind and its reasoning abilities constitute a glorious gift, granted so that we might know a greater mind from which ours is derived.

Many will regard all this as dangerous, a line of thought that threatens to drag our culture back into a benighted age of superstition. But we do well to recall that it was Judeo-Christian theism that served as the unique ground for the birth of science, as noted by Steve Fuller, a sociologist of science at the University of Warwick in England. Although he does not identify as a Christian or a practitioner of any religion,[9] he emphasizes what is common knowledge among historians of science: The conviction that nature is the work of a divine artificer, and that humans are made in the image of this being, fueled the birth of science by motivating its founders to go searching for an underlying order they had reason to believe they could successfully uncover. Isaac Newton, Johannes Kepler, and other scientific pioneers built their quest for knowledge on the conviction that the universe was created, and thus the study of the cosmos provides the wondrous prospect of probing and beholding the Creator's mind.[10]

In his foreword to a major volume critiquing theistic evolution, Fuller went further. "The cumulative effect of the set of papers assembled in this volume," he wrote, "is to suggest that the 'God hypothesis' (or what philosophers call 'divine action') remains very much on the table as a scientific explanation for events in the history of life."[11]

There is also this: A growing body of empirical evidence suggests that a life of theistic faith tends to produce attitudes and life habits that maintain or even improve one's epigenetic health. A healthy epigenome can then be passed as a legacy to future generations, a process that can feed a health-promoting cycle that gains momentum over time.

This is undoubtedly a controversial idea. What is the evidence for it? The topic would require a book in itself to treat adequately, but there is space here for a brief sketch of the accumulating data.

Epidemiologist Jeff Levin, who holds a distinguished chair at Baylor University, is author of many books on this topic. Two in particular summarize how theistic faith can positively impact many aspects of health: *God, Faith, and Health* (2002) and his Oxford University Press book *Religion and Medicine: A History of the Encounter Between Humanity's Two Greatest Institutions* (2020). He elucidates a series of findings suggesting diverse links between faith and health. In *Religion and Medicine* he summarizes his findings:

> If a reasonable conclusion can be drawn, it is simply this: for people of faith, religion can serve as a vital resource for coping with life's challenges, thus enhancing one's well-being.... The takeaway from the thousands of social, behavioral, biomedical, epidemiologic, and other clinical and population-health research studies is really quite straightforward and ought not be a lightning rod for controversy: namely, that "religious participation, on average, exhibits primary-preventive effects in well populations by an association with lower morbidity. No more, no less."[12]

As the "faith and health" literature has expanded enormously in recent decades, the body of empirical evidence for a positive connection is compelling enough that many medical schools have begun to include elective courses on faith and spirituality as life factors that affect human health. This development is squarely in the mainstream of medical education.[13]

Of course, the claim of an empirically verifiable connection between religious faith and human health has its vocal opponents. Most visible perhaps are the public intellectuals known as the new atheists, which includes Richard Dawkins along with his late colleagues Christopher Hitchens and Daniel Dennett. They have repeatedly insisted that traditional theistic faiths do serious damage to personal and social health.

Not all atheists agree. In 2008 *The Times* of London published an article by Matthew Parris, with the provocative title, "As an Atheist, I Truly Believe Africa Needs God." Parris, a British journalist and self-described atheist, shared his observations from living in Africa and traveling extensively from country to country across the continent. His conclusion from those experiences: "Education and training alone will not do. In Africa Christianity changes people's hearts. It brings a spiritual transformation. The rebirth is real. The change is good."[14]

Moral, behavioral, and social health—as affected by faith commitments—have become a major research focus in the case of one key social group that is a heavy burden on society: federal prison inmates. Their prior harmful patterns of behavior have been shown to be measurably changed, in aggregate, by a certain kind of personal faith experience. A series of peer-reviewed articles by sociologist Byron Johnson have tracked the tangible life changes among inmates who completed a Christian formation program instituted by Prison Fellowship. Many of Johnson's studies followed the life patterns of inmates once released from prison, to see what changes in behavior, if any, could be documented. The program sought to help inmates develop a life of faith focused on God as the central guiding and motivating force in their lives. Criminal quantitative studies by Johnson and others uncovered measurable and remarkable patterns of positive change.[15]

Johnson's series of articles caught the attention of those interested in addressing the tendency of prisoners to relapse into criminal behavior. Recidivism is a massive problem in the United States and other countries. When released inmates commit new offenses leading to reincarceration, it produces a cascade of costs to the individual and society that goes beyond the financial. Johnson found that the rate of recidivism was markedly lower among those who completed the Prison Fellowship study programs. In one major study recidivism dropped from 41 percent to 14 percent.[16] By living out a new life based on a love of God and loyal service to Him, and by cultivating behavior patterns marked by love and service to others, these former inmates showed a measurable transformation of life.

Figure 8.2. Byron Johnson of Baylor University is one of the world's top scholars on life impacts of personal faith and Bible study.

Johnson has published nearly a hundred research papers exploring the role of religious faith in combating social problems. He reviewed much of this evidence in his 2011 book *More God, Less Crime: Why Faith Matters and How It Could Matter More.*[17] The examples of beneficial changes to spiritual and physical health he explores there are compelling, and they hardly exhaust the evidence for positive life changes from theistic faith.[18]

At the same time, we ought to recall C. S. Lewis's cautionary note regarding religion, namely that "what good it will do you" is not how one should settle the debate between, for example, atheism and monotheism. Lewis sounded this warning in his impish essay "Man or Rabbit?," though there he set before the reader the scenario of someone attempting to resist theistic faith on the view that surely it was inessential from a practical perspective. In the essay, Lewis pictures a person who takes a purely pragmatic approach to the question of religious faith: "Can't you lead a good life without embracing a God-centered picture of reality?... I'm not interested in finding out whether the real universe is more what like the Christians say than what the Materialists say. All I'm interested in is leading a good life. I'm going to choose beliefs not because I think them true but because I find them helpful."[19] To this Lewis responds, "Now frankly, I find it hard to sympathize with this state of mind. One of the things that distinguishes man from the other animals is that he wants to know things, wants to find out what reality is like, simply for the sake of knowing. When that desire is completely quenched in anyone, I think he has become something less than human. As a matter of fact, I don't believe any of you have really lost that desire."[20]

So, yes, although we are convinced that theistic faith is good for you, we note the evidence for this not because we consider it dispositive but only to encourage agnostic readers to give the design hypothesis an open-minded hearing, for many of us have been told all manner of off-putting nonsense about the supposedly stultifying and warping effects of theistic belief. But our hope is that ultimately neither any ostensible convenience nor inconvenience of such theistic belief will play the primary role in your investigations of these matters. The admonition that rings out from Lewis's "Man or Rabbit?" is this: In considering the evidence for a designer of life, seek the truth, come what may. Any lesser counsel is surely unworthy of that most curious of animals, *Homo sapiens.*

ENDNOTES

1. Richard Francis, *Epigenetics: The Ultimate Mystery of Inheritance* (New York: W. W. Norton, 2011), xiii.

CHAPTER 1: DNA AND BEYOND

1. Bill Gates, Nathan Myhrvold, and Peter Rinearson, *The Road Ahead*, rev. ed. (Harmondsworth, UK: Viking Penguin Group, 1996), 228.

2. "Human Genome," *National Institutes of Health, National Center for Biotechnology Information*, accessed April 14, 2026, https://www.ncbi.nlm.nih.gov/gdv/.

3. Bill Clinton, "Remarks on the Completion of the First Survey of the Human Genome," *The American Presidency Project*, June 26, 2000, https://www.presidency.ucsb.edu/documents/remarks-the-completion -the-first-survey-the-human-genome.

4. "What They Said: Genome in Quotes," *BBC News: World Edition*, June 26, 2000, http://news.bbc.co.uk/2/hi/science/nature/807126.stm.

5. Nicholas Wade, "A Decade Later, Genetic Map Yields Few New Cures," *The New York Times*, June 12, 2010, https://www.nytimes.com/2010/06/13 /health/research/13genome.html.

6. Michael K. Skinner, "Role of Epigenetics in Developmental Biology and Transgenerational Inheritance," *Birth Defects Research Part C: Embryo Today: Reviews* 93, no. 1 (2011): 51–55, https://doi.org/10.1002/bdrc.20199.

7. Monya Baker, "Epigenome: Mapping in Motion," *Nature Methods* 7 (2010): 181–186, https://doi.org/10.1038/nmeth0310-181.

8. Gregory Matera and Zefeng Wang, "A Day in the Life of the Spliceosome," *Nature Reviews Molecular Cell Biology* 15 (March 11, 2014): 108–121, https://doi.org/10.1038/nrm3742.

9. Douglas L. Black, "Protein Diversity from Alternative Splicing: A Challenge for Bioinformatics and Post-Genome Biology," *Cell* 103, no. 3 (2000): 367, https://doi.org/10.1016/S0092-8674(00)00128-8.

10. Berkley E. Gryder, "The Designer Inference," accessed April 14, 2026, https://apologetics.org/wp-content/uploads/2024/07/TheDesignerInference atApologeticsInc.pdf.

11. For an example of the 90-percent claim, see S. Ohno, "An Argument for the Genetic Simplicity of Man and Other Mammals," *Journal of Human Evolution* 1, no. 6 (November 1972): 651–662, https://www.sciencedirect.com/science/article

/abs/pii/0047248472900115; and Alexander F. Palazzo and T. Ryan Gregory, "The Case for Junk DNA," *PLOS Genetics* (May 2014), https://pmc.ncbi.nlm.nih .gov/articles/PMC4014423/. For the 95-percent figure, see Philip Yam, "Talking Trash," *Scientific American* 272, no. 24 (March 1995); and Richard Dawkins, *The Greatest Show on Earth: The Evidence for Evolution* (The Free Press, 2009), 333. For the 99-percent figure, see Jack Lester King and Thomas H. Jukes, "Non-Darwinian Evolution," *Science* 164, no. 3881 (May 1969): 788–798, https:// www.science.org/doi/10.1126/science.164.3881.788; and Emily Mortola and Manyuan Long, "Turning Junk into Us: How Genes Are Born," *American Scientist* 109, no. 3 (May-June 2021), https://www.americanscientist.org/article/turning -junk-into-us-how-genes-are-born.

12. Ewan Birney, quoted in Ed Yong, "ENCODE: The Rough Guide to the Human Genome," *Discover*, September 5, 2012, https://www.discovermagazine.com /encode-the-rough-guide-to-the-human-genome-1703.

13. Richard Sternberg, Jonathan McLatchie, and Casey Luskin, "Here's a Far from Exhaustive (Yet Still Exhausting) List of Papers Discovering Function for 'Junk' DNA," *Science & Culture Today*, May 1, 2024, https://evolutionnews.org /2024/05/heres-a-far-from-exhaustive-yet-still-exhausting-list-of-papers -discovering-function-for-junk-dna/. See also Casey Luskin, "'Junk DNA' from Three Perspectives: Some Key Quotes," *Science & Culture Today*, May 2, 2024, https://evolutionnews.org/2024/05/junk-dna-from-three-perspectives-some -key-quotes/.

14. "Press Release: The Nobel Prize in Physiology or Medicine 2024," *Nobel Prize*, October 7, 2024, https://www.nobelprize.org/prizes/medicine/2024 /press-release/.

15. See for example B. W. Wright, M. P. Molloy, and P. R. Jaschke, "Overlapping Genes in Natural and Engineered Genomes," *Nature Reviews Genetics* 23 (2022): 154–168, https://doi.org/10.1038/s41576-021-00417-w. We could cite here several dozen other peer-reviewed papers and summaries that followed in the wake of the ENCODE results. For more on ENCODE, see "New Findings Challenge Established Views on Human Genome: ENCODE Research Consortium Uncovers Surprises Related to Organization and Function of Human Genetic Blueprint," *National Institutes of Health*, National Human Genome Research Institute, June 13, 2007, http://genome.gov/25521554.

16. Berkley Gryder, "Diving into the Mystery of Epigenetics," interview, *The Universe Next Door*, July 21, 2022, podcast, audio, https://theuniversenextdoor.podbean .com/e/diving-into-the-mystery-of-epigenetics-dr-berkley-gryder-dr-tom -woodward/.

17. It is much more complicated than this. See Chapter 3 for more details.

18. Martin I. Lind and Foteini Spagopoulou, "Evolutionary Consequences of Epigenetic Inheritance," *Heredity* 121 (2018): 205–209, https://doi.org/10.1038 /s41437-018-0113-y.

19. Lars Olov Bygren et al., "Paternal Grandparental Exposure to Crop Failure or Surfeit During a Childhood Slow Growth Period: Epigenetic Marks on Grandchildren's Growth, Glucoregulatory and Stress Genes," *bioRxiv* 215467 (December 13, 2018), https://doi.org/10.1101/215467. See also "Epigenetic Diseases and Their Causes and Symptoms," *Nature Scitable*, accessed April 14, 2026, https://www.nature.com/scitable/content/epigenetic-diseases-and-their -causes-and-symptoms-37397/ .

20. John Cloud, "Why Your DNA Isn't Your Destiny," *Time*, January 6, 2010, https://time.com/archive/6690329/why-your-dna-isnt-your-destiny/.

21. Paula M Lorenzo et al., "Epigenetic Effects of Healthy Foods and Lifestyle Habits from the Southern European Atlantic Diet Pattern: A Narrative Review," *Advances in Nutrition* 13, no. 5 (September 2022): 1725–1747, https://doi.org /10.1093/advances/nmac038.

22. Peter Lipton, "Inference to the Best Explanation," in W. H. Newton-Smith, ed., *A Companion to the Philosophy of Science* (Blackwell, 2000), 184–193, https://www. hps.cam.ac.uk/files/lipton-inference.pdf. See also Peter Lipton, *Inference to the Best Explanation*, 2nd ed. (London: Routledge, 2004). Stephen Meyer was able to meet Lipton and discuss inference to the best explanation (IBE) with him while studying for his Cambridge PhD in the early 1990s, and he provides a concise summary of how this approach came to inform his own investigation into the origin of life enigma. See Stephen C. Meyer, *Signature in the Cell: DNA and the Evidence for Intelligent Design* (New York: Harper One, 2009), 154–159.

23. See Carl Zimmer, "Scientists Seek to Update Evolution," *Quanta Magazine*, November 22, 2016, https://www.quantamagazine.org/scientists-seek-to-update -evolution-20161122/. Republished in *The Atlantic* as "The Biologists Who Want to Overhaul Evolution," November 28, 2016. See also "Why the Royal Society Meeting Mattered, In a Nutshell," *Science & Culture Today*, December 5, 2016, https://scienceandculture.com/2016/12/why_the_royal_s/.

24. Stephen Buranyi, "Do We Need a New Theory of Evolution?," *The Guardian*, June 28, 2022, https://www.theguardian.com/science/2022/jun/28/do-we-need -a-new-theory-of-evolution.

25. In addition to Buranyi's *Guardian* article cited above, see also Kevin N. Laland et al., "The Extended Evolutionary Synthesis: Its Structure, Assumptions and Predictions," *Proceedings of the Royal Society B: Biological Sciences* 282, no. 1813 (August 2015), https://doi.org/10.1098/rspb.2015.1019.

26. See citations to thirty-six peer-reviewed research articles at "Peer-Reviewed Articles Supporting Intelligent Design," *Discovery Institute*, Center for Science & Culture, accessed March 31, 2026, https://www.discovery.org/id/peer-review/. This is a subset of more than 250 such peer-reviewed papers, cited at "List of Peer-Reviewed and Mainstream Scientific Publications Supporting Intelligent Design," *Discovery Institute*, Center for Science & Culture, updated May 2024, https:// www.discovery.org/m/securepdfs/2024/05/Peer-Reviewed-and-Mainstream -Articles-Page-Update-May-2024_FinalPDF.pdf.

27. Michael J. Behe, "Train Wreck of a Review: A Response to Lenski et al. in *Science*," *Science & Culture Today*, February 14, 2019, https://evolutionnews. org/2019/02/train-wreck-of-a-review-a-response-to-lenski-et-al-in-science/. See also Behe's four additional responses to further criticism from Lenski following the "train wreck" review: Michael J. Behe, *A Mousetrap for Darwin* (Seattle: Discovery Institute Press, 2020), 410–441.

28. ID critic H. Allen Orr describes cells as possessing "staggering complexity" in "Devolution: Why Intelligent Design Isn't," *The New Yorker*, May 23, 2005, https://www.newyorker.com/magazine/2005/05/30/devolution-2. Michael Denton describes the cell as "infinitely complex" in *The Miracle of the Cell* (Seattle: Discovery Institute Press, 2020), 16.

29. John F. Kennedy, "Address at Rice University on the Nation's Space Effort," September 12, 1962, *John F. Kennedy Presidential Library and Museum*,

https://www.jfklibrary.org/archives/other-resources/john-f-kennedy-speeches
/rice-university-19620912.

Chapter 2: What Darwin Didn't Know About the Cell

1. Michael J. Behe, *Darwin's Black Box: The Biochemical Challenge to Evolution* (New York: Free Press, 2006), 255.

2. For those unfamiliar with the various parts of the DNA molecule, you may find helpful a demonstration that employs a hand-held model of DNA at https://dnaandbeyond.org/our-product/. The two-minute demonstration shows how the model functions with its twenty-one rungs and embedded magnets, complete with color-coded mnemonics for memorizing the different parts of the DNA molecule. A longer, ten-foot model of an RNA gene with seventy-five rungs is also at that same website, side by side with a model of the protein it produces.

3. Chad M. Kurylo et al., "Genome Sequence and Analysis of *Escherichia coli* MRE600, a Colicinogenic, Nonmotile Strain that Lacks RNase I and the Type I Methyltransferase, EcoKI," *Genome Biology and Evolution* 8, no. 3 (2016): 742–752, https://doi.org/10.1093/gbe/evw008.

4. Yesenia Cevallos et al., "A Brief Review on DNA Storage, Compression, and Digitalization," *Nano Communication Networks* 31 (2022): 100391, https://doi.org/10.1016/j.nancom.2021.100391.

5. Xingyu Liao et al., "Repetitive DNA Sequence Detection and Its Role in the Human Genome," *Communications Biology* 6, no. 954 (September 19, 2023), https://pmc.ncbi.nlm.nih.gov/articles/PMC10509279/.

6. Geoffrey M. Cooper, "The Nuclear Envelope and Traffic Between the Nucleus and Cytoplasm," in *The Cell: A Molecular Approach*, 2nd ed. (Sunderland, MA: Sinauer Associates, 2000).

7. Richard C. Francis, *Epigenetics: How Environment Shapes Our Genes* (New York: W. W. Norton, 2011), 21.

8. Sections that are removed are called introns. Sections that are retained and knit together are called exons. For an animation of the spliceosome at work in removing introns from an mRNA molecule, see "RNA Splicing," DNA Learning Center, *YouTube*, April 13, 2012, video, 1:37, https://www.youtube.com/watch?v=aVgwr0QpYNE.

9. Another barrel-shaped machine, a chaperone protein, often jumps into the action and works like a mechanized "dressing room," helping the protein fold in its private compartment.

10. For an animated rendering of the kinesin in action, see "The Workhorse of the Cell: Kinesin," Discovery Science, *YouTube*, May 14, 2014, video, 3:32, https://www.youtube.com/watch?v=gbycQf1TbM08&t=10s.

11. Michael Denton, *Evolution: A Theory in Crisis* (Chevy Chase, MA: Adler & Adler, 1986), 250.

12. Brian Miller and Casey Luskin, "Defending Douglas Axe on the Rarity of Protein Folds," *Science & Culture Today*, November 29, 2023, https://scienceandculture.com/2023/11/defending-douglas-axe-on-the-rarity-of-protein-folds/.

13. Douglas D. Axe, "Estimating the Prevalence of Protein Sequences Adopting Functional Enzyme Folds," *Journal of Molecular Biology* 341 (2004): 1295–1315, https://doi.org/10.1016/j.jmb.2004.06.058.

14. Susan Fields and Mark Johnston, "Proteins Are the Workhorses of the Cell: Misdiagnosis of a Metabolic Malady," in *Genetic Twists of Fate* (Cambridge, MA: MIT Press, 2010), chap. 3, https://doi.org/10.7551/mitpress/8709.001.0001. On page 23, the authors explain the "workhorse" analogy this way: "While DNA gets all the glory, proteins do all the heavy lifting. Proteins are the tiny machines that carry out nearly every cellular process, working in conjunction with other constituents of the cell to keep it alive and carry out its functions. The proteins in these machines are like gears and flywheels and valves: they fit together with exquisite precision and act in synchrony to carry out a specific cellular task."

15. Anne-Ruxandra Carvunis et al. discuss the importance of mapping the reference interactome. "To fully map the reference interactome, it is operationally helpful to go beyond binary protein interactions and identify protein complexes within cells," they comment. "Protein complexes typically contain five to six different proteins, within a wide range from two to hundreds in a variety of stoichiometries." "Interactome Networks" in *Handbook of Systems Biology: Concepts and Insights*, eds. Marian Walhout, Marc Vidal, and Job Dekker (Waltham, MA: Academic Press, 2012), 45–64.

16. Safa Al-Amrani et al., "Proteomics: Concepts and Applications in Human Medicine," *World Journal of Biological Chemistry* 12, no. 5 (September 27, 2021): 57–69, https://doi.org/10.4331/wjbc.v12.i5.57.

17. Dariusz Plewczyński and Krzysztof Ginalski, "The Interactome: Predicting the Protein-Protein Interactions in Cells," *Cell Molecular Biology Letters* 14, no. 1 (2009): 1–22, https://doi.org/10.2478/s11658-008-0024-7.

18. Or more formally: "The central dogma of molecular biology is a theory stating that genetic information flows only in one direction, from DNA, to RNA, to protein, or RNA directly to protein." In "The Central Dogma of Molecular Biology," *National Institutes of Health*, National Human Genome Research Institute, accessed April 1, 2026, https://www.genome.gov/genetics-glossary/Central -Dogma#:~:text=Definition&text=Central%20dogma.,or%20RNA%20 directly%20to%20protein.

19. Stephen R. Quake, "The Cellular Dogma," *Cell* 187 (November 14, 2024): 6421, https://www.cell.com/cell/fulltext/S0092-8674(24)01211-X.

Chapter 3: A Voyage Through the Cell

1. Jan Sapp, *Beyond the Gene: Cytoplasmic Inheritance and the Struggle for Authority in Genetics* (New York: Oxford University Press, 1987), 89.

2. Paul Peixoto et al., "From 1957 to Nowadays: A Brief History of Epigenetics," *International Journal of Molecular Science* 21 (October 14, 2020), https://pubmed.ncbi.nlm.nih.gov/33066397/.

3. Stanford University, faculty page for Stephen Quake, https://med.stanford.edu /profiles/stephen-quake.

4. Stephen R. Quake, "The Cellular Dogma," *Cell* 187 (November 14, 2024): 6421, https://www.cell.com/cell/fulltext/S0092-8674(24)01211-X.

5. Quake, "The Cellular Dogma," 6421.

6. Quake, "The Cellular Dogma," 6422.

7. Quake, "The Cellular Dogma," 6422.

8. For a deeper dive into newly discovered exotic modifications to histones, a great place to start is a September 2024 review article in *Nature* where the authors cite nine units that serve as histone modifiers beyond methyl groups, phosphate groups, and acetyl groups: *lactylation, citrullination, crotonylation, succinylation, SUMOylation, propionylation, butyrylation, 2-hydroxyisobutyrylation,* and *2-hydroxybutyrylation*. Weiyi Yao, Xinting Hu, and Xin Wang, "Crossing Epigenetic Frontiers: The Intersection of Novel Histone Modifications and Diseases," *Signal Transduction and Targeted Therapy* 9, no. 232 (September 16, 2024), https://www.nature.com/articles/s41392-024-01918-w.

9. Some transcription factors, called pioneer transcription factors, can destabilize/open up closed chromatin.

10. Jean-Pierre Issa, "Epigenetic Therapy," interview by Sarah Holt, *NOVA Online*, August 2007, www.pbs.org/wgbh/nova/genes/issa.html. The interview was conducted on January 8, 2007.

11. David Cheishvili, Lisa Boureau, and Moshe Szyf, "DNA Demethylation and Invasive Cancer: Implications for Therapeutics," *British Journal of Pharmacology* (August 18, 2014), https://doi.org/10.1111/bph.12885.

12. For a helpful overview of oncogenes and tumor suppressor genes, see "Oncogenes, Tumor Suppressor Genes, and Cancer," American Cancer Society, accessed April 1, 2026, https://www.cancer.org/cancer/understanding-cancer/genes-and-cancer/oncogenes-tumor-suppressor-genes.html.

13. Atsuya Nishiyama and Makoto Nakanishi, "Navigating the DNA Methylation Landscape of Cancer," *Trends in Genetics* 37, no. 11 (June 10, 2021): 1012–1027, https://doi.org/10.1016/j.tig.2021.05.002.

CHAPTER 4: MYSTERIES OF THE ZYGOTE CODE

1. Kayleigh J. A. Orchard et al., "Characterization of Histone Modifications in Late-Stage Rotator Cuff Tendinopathy," Genes 14, no. 2 (February 15, 2023): 496, https://doi.org/10.3390/genes14020496.

2. The ENCODE Project Consortium, "A User's Guide to the Encyclopedia of DNA Elements (ENCODE)," *PLOS Biology* 9, no. 4 (April 19, 2011): e1001046, https://doi.org/10.1371/journal.pbio.1001046.

3. The NIH began the Roadmap Epigenomics Mapping Consortium in 2008 with the goal of producing "a public resource of human epigenomic data to catalyze basic biology and disease-oriented research." Three years after the flurry of papers in 2012, the consortium released "Integrative Analysis of 111 Reference Human Epigenomes," *Nature* 518 (February 18, 2015): 317–330, https://doi.org/10.1038/nature14248, fulfilling the consortium's goal. The consortium integrated information and annotated regulatory elements across 127 reference epigenomes, sixteen of which were part of the ENCODE project. For an interim report about the Epigenome Roadmap, see Kerri Smith, "Epigenome: The Symphony in Your Cells," *Nature* (February 15, 2015), https://www.nature.com/articles/nature.2015.16955.

4. Nicholas Wade, "From One Genome, Many Types of Cells. But How?," *The New York Times*, February 23, 2009, https://www.nytimes.com/2009/02/24/science /24chromatin.html.

5. In his opening paragraph, Wade writes: "Somehow each of the 200 different kinds of cells in the human body... must be reading off a different set of instructions written into the DNA." By using the prepositional phrase "into the DNA," Wade means that each cell reads from a different subset of instructions. The context makes this clear. We have reworded Wade's "into the DNA" as "above and beyond" the DNA, because the markers that tell each cell which script lines to read (and which ones are off limits) are not all themselves part of the DNA script.

6. Wade, "From One Genome."

7. In one, Wells underlined the importance of understanding and correcting serious scientific problems in typical textbook presentations of the famous drawings of embryos by nineteenth-century embryologist Ernst Haeckel. Haeckel's drawings, though revealed as fraudulent in the late 1800s, were presented to unsuspecting biology students for over a hundred years. Haeckel's work served as an icon in many introductory biology textbooks, apparent visual proof of Darwinian evolution. See Jonathan Wells, "Haeckel's Embryos and Evolution: Setting the Record Straight," *American Biology Teacher* 61 (May 1999): 345–349. Wells's *Icons of Evolution: Why Much of What We Teach About Evolution Is Wrong* (2000) trained a light on this and a wider pattern of misinformation in textbook discussions of evolution by detailing ten such misleading icons. He then updated and expanded the list of common icons in his 2017 sequel *Zombie Science: More Icons of Evolution*. His 2011 book, *The Myth of Junk DNA*, debunked a relatively recent icon of evolutionary theory, the idea that the vast majority of our DNA is useless junk left over from the process of blind evolution.

8. Jonathan Wells, "Membrane Patterns Carry Ontogenetic Information That Is Specified Independently of DNA," *Bio-Complexity* (May 22, 2014), https://bio-complexity.org/ojs/index.php/main/article/view/BIO-C.2014.2 /BIO-C.2014.2.

9. ZP3 (Zona Pellucida protein 3) is a crucial protein involved in fertilization. It serves as a receptor for sperm cells, binding to specific proteins on the sperm surface. This binding triggers the acrosome reaction, releasing enzymes that allow the sperm to penetrate the zona pellucida and fertilize the egg. Geoffrey M. Cooper and Robert E. Hausman, *The Cell: A Molecular Approach*, 7th ed. (Oxford, UK: Oxford University Press, 2014), 345.

10. Hans-Joachim Gabius and Jürgen Roth, "An Introduction to the Sugar Code," *Histochemistry and Cell Biology* 147 (December 14, 2016): 111–117, https://doi.org/10.1007/s00418-016-1521-9; Hans-Joachim Gabius, "Biological Information Transfer Beyond the Genetic Code: The Sugar Code," *Naturwissenschaften* 87, no. 3 (March 2000): 109, https://doi.org/10.1007 /s001140050687; Hans-Joachim Gabius et al., "Chemical Biology of the Sugar Code," *ChemBioChem* 5 (May 26, 2004): 741, https://doi.org/10.1002/ cbic.200300753. A 2018 article at *The Conversation* emphasizes how important the sugar code is. Yes, proteins are rightly considered the workhorses of the cellular realm, fulfilling numerous functions essential for life. But, as the authors explain, "how a protein behaves often depends on what glycans are attached to it. In other words, these sugar molecules can greatly influence how our proteins do their work,

and even how our cells will respond to stimuli. For example, if you change a few glycans on the outside of a cell, it might trigger that cell to migrate to a different location in our body." Emanuel Maverakis, Calito Lebrilla, and Jenny Wang, "Cracking the Sugar Code: Why the 'Glycome' Is the next Big Thing in Health and Medicine," *The Conversation*, August 28, 2018, https://theconversation .com/cracking-the-sugar-code-why-the-glycome-is-the-next-big-thing-in -health-and-medicine-97750.

11. Janine Beisson and T. M. Sonneborn, "Cytoplasmic Inheritance of the Organization of the Cell Cortex in Paramecium Aurelia," *PNAS* 53, no. 2 (February 15, 1965): 275–282, https://doi.org/10.1073/pnas.53.2.275.

12. "Cortical inheritance is a process that involves the spatial and structural organization of cellular components during cell division, particularly in the context of fertilization and early embryonic development," explains biologist Jonathan Wells. "It encompasses mechanisms such as cortical rotation and the cortical reaction, which are crucial for establishing cellular polarity and preventing polyspermy. In contrast, genetic inheritance pertains specifically to the transmission of genetic information encoded in DNA from one generation to the next. While both processes are essential for proper development, they operate through fundamentally different mechanisms. Cortical inheritance does not involve the transfer of genetic material; instead, it focuses on the distribution of cytoplasmic components and organelles that influence cellular behavior post-fertilization." From personal communication, Jonathan Wells. See also Janine Beisson, who masterfully investigates cortical inheritance in "Preformed Cell Structure and Cell Heredity," *Prion* 2, no. 1 (January-March 2008): 1–8, https://doi.org/10.4161/pri.2.1.5063.

13. Stephen F. Ng and Joseph Frankel, "180 Degree Rotation of Ciliary Rows and Its Morphogenetic Implications in Tetrahymena Pyriformis," *PNAS* 74, no. 3 (March 1977): 1118, http://www.pnas.org/content/74/3/1115.full.pdf+html.

Chapter 5: Epigenetics and Health

1. See, for example, Shikhar Sharma, Theresa K. Kelly, and Peter A Jones, "Epigenetics in Cancer," *Carcinogenesis* 31, no. 1 (January 2010): 27–36, https://doi.org/10.1093/carcin/bgp220.

2. Wanlin Dai et al., "Epigenetics-Targeted Drugs: Current Paradigms and Future Challenges," *Signal Transduction and Targeted Therapy* 9, Article 332 (November 2024), https://doi.org/10.1038/s41392-024-02039-0.

3. G. Kaati, L. O. Bygren, and S. Edvinsson, "Cardiovascular and Diabetes Mortality Determined by Nutrition During Parents' and Grandparents' Slow Growth Period," *European Journal of Human Genetics* 10, no. 11 (November 2002): 682–688, https://doi.org/10.1038/sj.ejhg.5200859.

4. R. A. Waterland and R. L. Jirtle, "Transposable Elements: Targets for Early Nutritional Effects on Epigenetic Gene Regulation," *Molecular and Cellular Biology* 23, no. 15 (March 27, 2003): 5293–5300, https://doi.org/10.1128 /MCB.23.15.5293-5300.2003.

5. Randy Jirtle, "Ask the Expert," *Nova*, August 2, 2007, https://www.pbs.org/wgbh /nova/genes/expert.html. Jirtle cites research by Dana C. Dolinoy, Dale Huang, and himself, "Maternal Nutrient Supplementation Counteracts Bisphenol

A-Induced DNA Hypomethylation in Early Development," *PNAS* 104, no. 32 (August 7, 2007): 13056–13061, http://www.pnas.org/content/104/32/13056.full .pdf+html.

6. Randy Jirtle quoted in Ethan Watters, "DNA Is Not Destiny: The New Science of Epigenetics," *Discover*, November 21, 2006, https://www.discovermagazine.com /dna-is-not-destiny-the-new-science-of-epigenetics-1746.

7. Randy Jirtle quoted in Bob Weinhold, "Epigenetics: The Science of Change," *Environmental Health Perspectives* 114, no. 3 (March 2006): A160–A167, https:// pmc.ncbi.nlm.nih.gov/articles/PMC1392256/. For a fascinating account of Jirtle's research, see John Cloud, "Why Your DNA Isn't Your Destiny," *Time*, January 6, 2010, 50. See also Jirtle's Q&A at *Nova: Ghost in Your Genes*, conducted August 2 and November 1 of 2007, https://www.pbs.org/wgbh/nova/genes/expert.html.

8. Sang-Woon Choi and Simonetta Friso, "Epigenetics: A New Bridge Between Nutrition and Health," *Advances in Nutrition* 1, no. 1 (November 2010): 8–16, https://doi.org/10.3945/an.110.1004.

9. Dina Drits-Esser et al., "Beyond the Central Dogma: Bringing Epigenetics into the Classroom," *The American Biology Teacher* 76, no. 6 (August 2014): 365–369, https://doi.org/10.1525/abt.2014.76.6.3.

10. Jenny Lee et al., "An Epigenetics-Based, Lifestyle Medicine-Driven Approach to Stress Management for Primary Patient Care: Implications for Medical Education," *American Journal of Lifestyle Medicine* 14, no. 3 (May 9, 2019): 294–303, https://doi.org/10.1177/1559827619847436.

11. M. Reza Sailani et al., "Lifelong Physical Activity Is Associated with Promoter Hypomethylation of Genes Involved in Metabolism, Myogenesis, Contractile Properties, and Oxidative Stress Resistance in Aged Human Skeletal Muscle," *Scientific Reports* 9 (2019): 3272, https://doi.org/10.1038/s41598-018-37895-8.

12. Zidian Xie et al., "Perspectives on Epigenetics Alterations Associated with Smoking and Vaping," *Function* 2, no. 3 (April 23, 2021): zqab022, https://doi.org /10.1093/function/zqab022.

13. *The Health Consequences of Smoking—50 Years of Progress: A Report of the Surgeon General* (Atlanta, GA: Centers for Disease Control and Prevention, 2014), https://www.ncbi.nlm.nih.gov/books/NBK179276/.

14. Michael J. Meaney and Moshe Szyf, "Environmental Programming of Stress Responses Through DNA Methylation: Life at the Interface Between a Dynamic Environment and a Fixed Genome," *Dialogues in Clinical Neuroscience* 7, no. 2 (2005): 103–123, https://doi.org/10.31887/DCNS.2005.7.2/mmeaney.

15. "Learning Without Learning," *The Economist*, September 21, 2006, https://www.economist.com/science-and-technology/2006/09/21 /learning-without-learning.

16. The mechanics of how the epigenetic changes were triggered by the grooming and nurturing behavior is complex. These processes, along with a powerful biological trigger—a protein called NGFI-A—are summarized in "Learning Without Learning."

17. "Learning Without Learning."

18. "Acute Stress Leaves Epigenetic Marks on the Hippocampus," *The Rockefeller University*, November 23, 2009, https://www.rockefeller.edu/news/1185-acute -stress-leaves-epigenetic-marks-on-the-hippocampus/.

19. "Acute Stress Leaves Epigenetic Marks." The research study that the article reports on is Richard G. Hunter et al., "Regulation of Hippocampal H3 Histone Methylation by Acute and Chronic Stress," *PNAS* (December 8, 2009). See also Richard G. Hunter et al., "Acute Stress and Hippocampal Histone H3 Lysine 9 Trimethylation, a Retrotransposon Silencing Response," *PNAS* 109, no. 43 (October 4, 2012): 17657–17662, https://doi.org/10.1073/pnas.1215810109.

20. Richard S. Lee et al., "Chronic Corticosterone Exposure Increases Expression and Decreases Deoxyribonucleic Acid Methylation of Fkbp5 in Mice," *Endocrinology* 151, no. 9 (September 1, 2010): 4332–4343, https://doi.org/10.1210/en.2010-0225.

21. "Chronic Stress May Cause Long-Lasting Changes," *JHU Gazette*, October 25, 2010, http://gazette.jhu.edu/2010/10/25/chronic-stress-may-cause-long-lasting-changes.

22. "Chronic Stress May Cause Long-Lasting Changes."

23. Harrison Wein, "Stress Hormone Causes Epigenetic Changes," *NIH Research Matters*, September 27, 2010, https://web.archive.org/web/20190716052015/https://www.nih.gov/news-events/nih-research-matters/stress-hormone-causes-epigenetic-changes.

24. James Potash quoted in "Chronic Stress May Cause Long-lasting Changes."

25. Ruth Williams, "A Trip to the Gym Alters DNA," *Nature* (March 6, 2012), https://doi.org/10.1038/nature.2012.10176. Williams was reporting on research published by Romain Barrès et al., "Acute Exercise Remodels Promoter Methylation in Human Skeletal Muscle," *Cell Metabolism* 15, no. 3 (March 7, 2012): 405–411, https://doi.org/10.1016/j.cmet.2012.01.001.

26. Julio Plaza-Diaz et al., "Impact of Physical Activity and Exercise on the Epigenome in Skeletal Muscle and Effects on Systemic Metabolism," *Biomedicines* 10, no. 1 (January 10, 2022): 126, https://doi.org/10.3390/biomedicines10010126.

Chapter 6: Genomic Complexity: How Deep Does It Go?

1. Erika Check Hayden, "Human Genome at Ten: Life Is Complicated," *Nature* 464 (April 1, 2010): 664, https://doi.org/10.1038/464664a. "'When we started out, the idea was that signalling pathways were fairly simple and linear,' says Tony Pawson, a cell biologist at the University of Toronto in Ontario. 'Now, we appreciate that the signalling information in cells is organized through networks of information rather than simple discrete pathways. It's infinitely more complex.'"

2. See the important overview by Magdalena Koziol and John L. Rinn, "RNA Traffic Control of Chromatin Complexes," *Current Opinion in Genetics and Development* 20, no. 2 (April 2010): 142–148, https://doi.org/10.1016/j.gde.2010.03.003.

3. Mitchell Guttman et al., "lincRNAs Act in the Circuitry Controlling Pluripotency and Differentiation," *Nature* 477 (2011): 295–300, https://doi.org/10.1038/nature10398.

4. Quoted in "New Roles Emerge for Non-Coding RNAs in Directing Embryonic Development," *ScienceDaily*, August 29, 2011, http://www.sciencedaily.com/releases/2011/08/110828141048.htm.

5. Aaron David Goldman and Laura F. Landweber, "What Is a Genome?," *PLOS Genetics* 12, no 7 (July 21, 2016): https://doi.org/10.1371/journal.pgen.1006181.

6. Richard Sternberg expressed this in talks given at a conference on epigenetics we organized in Clearwater, Florida, held in February 2012. The exact wording is found in a 2014 upload of PowerPoint slides he used, accessed June 27, 2012, https://www.slideserve.com/jenski/richard-v-sternberg. Sternberg's ideas regarding the "immaterial gene" have been beautifully teased out by David Klinghoffer in *Plato's Revenge: The New Science of the Immaterial Genome* (Seattle: Discovery Institute Press, 2025).

7. Rahul Jagadeesan et al., "Dynamics of Bacterial Operons During Genome-Wide Stresses Is Influenced by Premature Terminations and Internal Promoters," *Science Advances* 11, no. 20 (May 16, 2025), https://www.science.org/doi/10.1126/sciadv.adl3570.

8. See, for example, Wen-Yu Chung et al., "A First Look at ARFome: Dual-Coding Genes in Mammalian Genomes," *PLoS Computational Biology* 3, no. 5 (May 18, 2007), https://doi.org/10.1371/journal.pcbi.0030091. The authors comment: "Coding of multiple proteins by overlapping reading frames is not a feature one would associate with eukaryotic genes. Indeed, codependency between codons of overlapping protein-coding regions imposes a unique set of evolutionary constraints, making it a costly arrangement. Yet in cases of tightly coexpressed interacting proteins, dual coding may be advantageous. Here we show that. although dual coding is nearly impossible by chance, a number of human transcripts contain overlapping coding regions. Using newly developed statistical techniques, we identified 40 candidate genes with evolutionarily conserved overlapping coding regions. Because our approach is conservative, we expect mammals to possess more dual-coding genes. Our results emphasize that the skepticism surrounding eukaryotic dual coding is unwarranted: rather than being artifacts, overlapping reading frames are often hallmarks of fascinating biology."

9. Gary Taubes, "The Sea Change That's Challenging Biology's Central Dogma," *Discover*, November 3, 2009, https://www.discovermagazine.com/the-sea-change-thats-challenging-biologys-central-dogma-12937.

10. Elizabeth Pennisi, "'Dark Proteome' Survey Reveals Thousands of New Human Genes," *Science* (November 25, 2024), https://www.science.org/content/article/dark-proteome-survey-reveals-thousands-new-human-genes

11. Dietmar Schmucker et al., "Drosophila Dscam Is an Axon Guidance Receptor Exhibiting Extraordinary Molecular Diversity," *Cell* 101 (June 9, 2000): 672, https://www.cell.com/cell/fulltext/S0092-8674(00)80878-8. They comment, "In the course of the analysis of Dscam cDNAs, we uncovered extraordinary sequence diversity in the extracellular region. Through splicing of alternative exons encoding three different Ig domains and the transmembrane domain, more than 38,000 proteins may be generated. These proteins would have an identical architecture but differ in their combination of the four variable domains. We also show that Dscam is required in other neurons in the developing central nervous system for the establishment of axon pathways. This raises the intriguing possibility that diverse forms of a single protein may contribute to specifying the precise connections made by many different types of neurons."

12. James B. Brown et al., "Diversity and Dynamics of the Drosophila Transcriptome," *Nature* 512 (March 16, 2014): 393–399, https://doi.org/10.1038/nature12962. See also Jonatha M. Gott, "Expanding Genome Capacity via RNA Editing," *Comptes Rendus Biologies* 326, no. 10–11 (October 2003): 901–908,

https://comptes-rendus.academie-sciences.fr/biologies/item/10.1016/j.crvi
.2003.09.004.pdf.

13. Amartya Sanyal et al., "The Long-Range Interaction Landscape of Gene
Promoters," *Nature* 489 (2012): 109–113, https://doi.org/10.1038/nature11279.

14. Daniel Panne, Tom Maniatis, and Stephen C. Harrison, *"An Atomic Model of the
Interferon-β Enhanceosome,"* *Cell* 129, no. 6 (June 15, 2007): 1119, DOI: 10.1016/j.
cell.2007.05.019. Internal references elided.

15. Anil Panigrahi and Bert W. O'Malley, "Mechanisms of Enhancer Action: The
Known and the Unknown," *Genome Biology* 22, article 108 (2021), https://doi
.org/10.1186/s13059-021-02322-1. Although enhancer sections of DNA are
always associated with the start point of a gene's sequence (as are promoters), they
generally are farther away than promoter sequences, sometimes hundreds or even
thousands of rungs away. In one amazing case the enhancer for a gene labeled
SHH is located a million rungs away.

16. Berkley Gryder, Gryder Lab, Case Western Reserve University, personal email
communication, February 2, 2025.

17. Hao Liu et al., "Three-Dimensional Genome Structure and Function," *MedComm*
4, no. 4 (July 8, 2023): 1, https://doi.org/10.1002/mco2.326.

18. Liu et al., "Three-Dimensional Genome," 1.

19. Zhijun Duan et al., "A Three-Dimensional Model of the Yeast Genome," *Nature*
465 (May 2, 2010): 363, https://doi.org/10.1038/nature08973. Internal references
removed.

20. Cristina Luiggi, "Genome Blossoms," *The Scientist* (November 1, 2010),
https://www.the-scientist.com/genome-blossoms-43002.

21. Luiggi, "Genome Blossoms."

22. Eric Werner, "What Makes Us Human? Or Why Aren't We Mice?," *PLoS Biology*
(May 27, 2009), https://journals.plos.org/plosbiology/article/comment?id=10.1371
/annotation/15134b58-7375-48e7-9553-42979b51bb5d, reader comment under
Deanna M. Church et al., "Lineage-Specific Biology Revealed by a Finished
Genome Assembly of the Mouse," *PLoS Biology* (May 26, 2009), https://doi.
org/10.1371/journal.pbio.1000112.

CHAPTER 7: SCRABBLING FOR AN ADEQUATE CAUSE

1. "Remarks by the President, Prime Minister Tony Blair of England (via Satellite),
Dr. Francis Collins… and Craig Venter… on the Completion of the First Survey
of the Entire Human Genome Project," June 26, 2000, *The White House*,
https://clintonwhitehouse3.archives.gov/WH/EOP/OSTP/html/00628_2.html.

2. For the portion of Collins's talk where he relates the significance of reading C. S.
Lewis's *Mere Christianity*, see minutes 6–15 of "My Journey from Atheism to
Christianity: Francis Collins at Caltech," Veritas Forum, *YouTube*, September 6,
2016, video, 27:52, https://www.youtube.com/watch?v=HaEQyNeaFZs.

3. This is the picture Francis Collins paints in *The Language of God: A Scientist
Presents Evidence for Belief* (New York: Free Press, 2006), 22–31, 57–84.

4. Collins, *The Language of God*, 181–195.

5. James Watson and Andrew Berry, *DNA: The Secret of Life* (New York: Alfred A
Knopf, 2003), 404.

6. Watson and Berry, *DNA*, 405.

7. William A. Dembski and Winston Ewert, *The Design Inference: Eliminating Chance Through Small Probabilities*, 2nd ed. (Seattle: Discovery Institute Press, 2023), 321–394.

8. Michael J. Behe, *Darwin's Black Box: The Biochemical Challenge to Evolution*, 10th anniversary edition (New York: Free Press, 2006), 193.

9. Michael J. Behe, *The Edge of Evolution: The Search for the Limits of Darwinism* (New York: Free Press, 2007).

10. Michael J. Behe, *Darwin Devolves: The New Science About DNA That Challenges Evolution* (New York: HarperOne, 2019).

11. Michael J. Behe, *A Mousetrap for Darwin: Michael J. Behe Answers His Critics* (Seattle: Discovery Institute Press, 2020). For a more concise response to the principal criticisms lodged against Behe's design arguments, see his appendix in *Darwin Devolves*.

12. Stephen C. Meyer, *Signature in the Cell: DNA and the Evidence for Intelligent Design* (New York: HarperOne, 2009); *Darwin's Doubt: The Explosive Origin of Animal Life and the Case for Intelligent Design* (New York: Harper One, 2013); *Return of the God Hypothesis: Three Scientific Discoveries That Reveal the Mind Behind the Universe* (New York: HarperOne, 2020).

13. For a review of the intense attacks on Behe and other ID scientists, see Thomas Woodward, *Darwin Strikes Back: Defending the Science of Intelligent Design* (Grand Rapids, MI: Baker, 2006), chaps. 3, 4, and 5. See also *Revolutionary*, Discovery Science, *YouTube*, September 11, 2017, video, 59:55, https://www.youtube.com/watch?v=7ToSEAj2V0s.

14. We should note in passing that this Royal Society meeting came fully three decades after Australian geneticist Michael Denton sent out much the same memo via his 1985 book *Evolution: A Theory in Crisis* (London: Burnett Books).

15. The comments by George Church and Mark McMenamin can be found in the front matter of Stephen Meyer's *Darwin's Doubt*.

16. See John W. Kimball, "Genome Sizes," *LibreTexts Biology*, accessed April 1, 2026, https://bio.libretexts.org/Bookshelves/Introductory_and_General_Biology/Biology_(Kimball)/05%3A_DNA/5.09%3A_Genome_Sizes. Kimball is author of *Biology*, 6th ed. (Dubuque, IA: Wm. C. Brown, 1994).

17. See for example, Douglas Axe's peer-reviewed work on proteins: "Extreme Functional Sensitivity to Conservative Amino Acid Changes on Enzyme Exteriors," *Journal of Molecular Biology* 301 (2000): 585–595, https://doi.org/10.1006/jmbi.2000.3997; "Estimating the Prevalence of Protein Sequences Adopting Functional Enzyme Folds," *Journal of Molecular Biology* 341 (2004): 1295–1315, https://doi.org/10.1016/j.jmb.2004.06.058; and "The Case Against a Darwinian Origin of Protein Folds," *BIO-Complexity* 2010, no. 1 (April 15, 2010), https://bio-complexity.org/ojs/index.php/main/article/view/bio-c.2010.1.

18. See, for instance, Francisco J. Blanco, Isabelle Angrand, and Luis Serrano, "Exploring the Conformational Properties of the Sequence Space Between Two Proteins with Different Folds: An Experimental Study," *Journal of Molecular Biology* 285, no. 2 (January 15, 1999): 741–753; and Shimon Bershtein, Michael Segal, Roy Bekerman, Nobuhiko Tokuriki, and Dan S. Tawfik, "Robustness-Epistasis Link Shapes the Fitness Landscape of a Randomly Drifting Protein," *Nature* 444 (2006): 929–932, https://doi.org/10.1038/nature05385. For a discussion of the work's relevance to Axe's research on protein folding, see Meyer, *Darwin's Doubt*, 197.

19. For a comprehensive list of biological codes, see Robert Prinz, "Code Biology Database—A List of Biological Codes," *International Society of Code Biology*, December 2022, https://codebiology.org/database.pdf. For more on biological codes, see Robert Prinz, "Biological Codes: A Field Guide for Code Hunters," *Biological Theory* 19 (July 6, 2023): 120–136, https://doi.org/10.1007/s13752 -023-00444-2.

20. Berkley E. Gryder, "The Designer Inference," accessed February 11, 2026, https://apologetics.org/wp-content/uploads/2024/07/TheDesignerInferenceat ApologeticsInc.pdf, 17.

21. Peppi Koivunen and Tuomas Laukka, "The TET Enzymes," *Cellular and Molecular Life Sciences* 75, no. 8 (November 28, 2017): 1339–1348, https://doi .org/10.1007/s00018-017-2721-8.

22. Pengfei Tian and Robert B. Best. 2017. "How Many Protein Sequences Fold to a Given Structure? A Coevolutionary Analysis," *Biophysical Journal* 113 (2017): 1719–1730, DOI: 10.1016/j.bpj.2017.08.039.

23. Axe, "Estimating the Prevalence."

24. Agnes K. M. Weiner et al., "Phylogenomics of the Epigenetic Toolkit Reveals Punctate Retention of Genes Across Eukaryotes," *Genome Biological Evolution* 12, no. 12 (December 2020): 2196, https://doi.org/10.1093/gbe/evaa198.

Chapter 8: Peering into the Known And Unknown

1. Paul Davies and Niels Henrik Gregersen, eds., *Information and the Nature of Reality: From Physics to Metaphysics* (Cambridge, UK: Cambridge University Press, 2014).

2. Ieva Rauluseviciute, Finn Drabløs, and Morten Beck Rye, "DNA Hypermethylation Associated with Upregulated Gene Expression in Prostate Cancer Demonstrates the Diversity of Epigenetic Regulation," *BMC Medical Genomics* 13, article 6 (2020), https://doi.org/10.1186/s12920-020-0657-6.

3. Rauluseviciute et al., "DNA Hypermethylation." See also "Scientists Map Epigenome of Human Stem Cells During Development," *ScienceDaily*, February 4, 2010, http://www.sciencedaily.com/releases/2010/02/100203141326.htm.

4. Mahmoud Alhosin, "Epigenetics Mechanisms of Honeybees: Secrets of Royal Jelly," *Epigenetic Insights* 16 (November 29, 2023), https://doi.org/10.1177 /25168657231213717. See also "Epigenetics: What Makes a Queen Bee?," *Nature* 468, no. 348 (2010), https://doi.org/10.1038/468348a.

5. Michael J. Behe, *Darwin Devolves: The New Science About DNA That Challenges Evolution* (New York: HarperOne, 2019).

6. See the helpful study by Hannah Ritchie, "How Many Species Are There?," *Our World in Data*, November 30, 2022, https://ourworldindata.org/how-many -species-are-there. Estimates range as high as 8.7 million, and with bacteria and archaea thrown in, some speculate that over a billion species exist, most of them still waiting to be discovered.

7. Richard Dawkins, *The Blind Watchmaker: Why the Evidence of Evolution Reveals a Universe Without Design* (New York: W. W. Norton, 1986), 1.

8. Dawkins, *The Blind Watchmaker*, 21.

9. Steve Fuller, personal email communication, November 14, 2024. In the exchange he described himself as "perhaps monotheistically inclined."

10. These ideas are found in many of Steve Fuller's books, especially those that probe the intelligent design controversy. See, for example, *Dissent over Descent* (London: Icon, 2008). He also has laid out these arguments in videos. See, for example, the dialog at "Why Are You Threatened by Intelligent Design?," The Renewed Mind, *YouTube*, September16, 2009, video, 7:05, filmed July 11, 2009 at Cambridge University, https://www.youtube.com/watch?v=T0yerBAqG9Y. For more on the connection between the rise of science and the Judeo-Christian worldview, see Peter Harrison, *The Territories of Science and Religion* (Chicago: University of Chicago Press, 2015).

11. Steve Fuller, foreword, *Theistic Evolution: A Scientific, Philosophical, and Theological Critique*, eds. J. P. Moreland et al. (Wheaton, Illinois: Crossway, 2017), 27.

12. Jeff Levin, *Religion and Medicine: A History of the Encounter Between Humanity's Two Greatest Institutions* (Oxford, UK: Oxford University Press, 2020), 107. The quote that concludes Levin's summary here is from Jeff Levin, Linda M. Chatters, and Robert Joseph Taylor, "Religious Factors in Health and Medical Care among Older Adults," *Southern Medical Journal* 99 (2006): 1168-1169.

13. See especially Harold G. Koenig, Tyler J. VanderWeele, and John R. Peteet, *Handbook of Religion and Health*, 3rd ed. (Oxford, UK: Oxford University Press, 2023).

14. Matthew Parris, "As an Atheist, I Truly Believe Africa Needs God," *The Times* (London), December 27, 2008, http://www.rootedinjesus.net/docs/Parris.pdf.

15. See Byron R. Johnson, David B. Larson, and Timothy G. Pitts, "Religious Programs, Institutional Adjustment, and Recidivism Among Former Inmates in Prison Fellowship Programs," *Justice Quarterly* 14, no. 1 (March 1997): 145–166; Byron R. Johnson and David B. Larson, "Linking Religion to the Mental and Physical Health of Inmates: A Literature Review and Research Note," *American Jails* 11, no. 4 (September/October 1997): 28–36; Byron R. Johnson, Sung Joon Jang, and Christopher Bader, "The Cumulative Advantage of Religiosity in Preventing Drug Use," *Journal of Drug* 38, no. 3 (July 2008): 771–798. See also the description of this methodology in M. L. Dantzker and Ronald D. Hunter, *Research Methods for Criminology and Criminal Justice: A Primer* (Boston, MA: Butterworth-Heinemann, 2000).

16. Byron R. Johnson, "Religious Programs and Recidivism Among Former Inmates in Prison Fellowship Programs: A Long-Term Follow-Up Study," *Justice Quarterly* 21, no. 2 (2004): 329–354.

17. Byron R. Johnson, *More God, Less Crime: Why Faith Matters and How It Could Matter More* (West Conshohocken, PA: Templeton Press, 2011).

18. An overview of Byron Johnson's research can be found at the website for Baylor University's Institute for Global Human Flourishing: https://humanflourishing. web.baylor.edu/person/dr-byron-johnson. His scholarship examines "ways in which religion impacts key behaviors like volunteerism, generosity, and purpose. These topics are covered in four recent books, *The Angola Prison Seminary* (2016), which evaluates the influence of a Bible College and inmate-led congregations on prisoners serving life sentences; *The Quest for Purpose: The Collegiate Search for a Meaningful Life* (2017), which examines the link between religion and finding purpose and meaning, and the subsequent link to academic integrity; *The Restorative Prison: Essays on Inmate Peer Ministry and Prosocial Corrections* (2021), which looks at the empirical evidence in support of the link between religion and

the emerging subfield of positive criminology; and *Objective Religion: Freedom, Politics, Secularization* (2023), which examines factors related to the importance and resilience of religion." We would add that current research in the health/religion interface involves both qualitative evidence as well as quantitative measures. The qualitative side of the evidence draws heavily on personal life stories, and social scientists today pay more attention than ever to such qualitative studies. We have done our own qualitative research on the impact of theistic faith among professors at Princeton University. In our investigation, we noted and video-recorded the positive changes that flowed from such belief and from related faith behavior in the lives of five Princeton professors who came to monotheistic belief by very different pathways. We focused on the area of spiritual health as it flows outward into positive change in the areas of intellectual vibrancy and curiosity, moral coherence, and social flourishing. The highlight of these investigations was a set of interviews that explored the life path of each professor. The interviews can be viewed at "Dr. Biagini and Princeton Chronicles," *Apologetics, Inc.*, https://apologetics.org/princeton. Much more needs to be done to research the life stories of those who went in the other direction—those who abandoned a formerly held belief in God as part of a turn to naturalism or scientific materialism.

19. C. S. Lewis, "Man or Rabbit?," *God in the Dock: Essays on Theology and Ethics* (New York: William B. Eerdmans, 1970), 108.

20. Lewis, "Man or Rabbit?," 108. Lewis goes on to lay out the priority of the question "Is it true?" over "What good will it do me?" He then returns to the "what good will it do me?" issue. He says this new theistic vista "will do you good—a great deal more good than you ever wanted or expected" (112). He concludes the essay with these memorable words: "The idea of reaching 'a good life' without Christ is based on a double error. Firstly, we cannot do it; and secondly, in setting up 'a good life' as our final goal, we have missed the very point of our existence. Morality is a mountain which we cannot climb by our own efforts; and if we could we should only perish in the ice and unbreathable air of the summit, lacking those wings with which the rest of the journey has to be accomplished. For it is from there that the real ascent begins. The ropes and axes are 'done away' and the rest is a matter of flying" (112–113).

ACKNOWLEDGMENTS

We are indebted to Discovery Institute and to so many who assisted us in our years of biological research and analysis. *Epigenetics and the Architect* is a substantially revised and updated version of our earlier book, *The Mysterious Epigenome: What Lies Beyond DNA*, from Kregel Publications. We appreciate the assistance of Jerry Kregel and Dennis Hillman in both crafting the original book and facilitating the transition to a wholly new edition.

We also want to thank Casey Luskin, Jonathan McLatchie, Emily Reeves, and our other peer reviewers. Accuracy is paramount, and while any errors that remain are strictly those of the authors, there are assuredly far fewer, thanks to their expert advice.

We appreciate the wise oversight of Jonathan Witt and all who helped in the editing process, and we salute the contribution of Guillermo Cuadra in researching recent scientific breakthroughs. In the task of gathering helpful illustrations, we are indebted to Joseph Condeelis and Illustra Media, along with so many others.

In addition, about two dozen scientists, most of whom teach and conduct research at major universities, were the backbone of this book. We appreciate their collegial spirit and time invested as they shared with us from their vast knowledge and wisdom. It illuminated our path each step of the way. Most of all, we acknowledge the sacrifice and patience of our wives as we toiled to produce and edit the final manuscript. *Soli Deo gloria.*

—Thomas Woodward and James P. Gills

Image Credits

Figure 1.1. A stretch of DNA wrapped around histones. Image by Joseph Condeelis.

Figure 1.2. Genetics and epigenetics compared. Table by Thomas Woodward.

Figure 2.1. Charles Darwin. Photograph by Henry Maull and John Fox. Frontispiece of Francis Darwin's *The Life and Letters of Charles Darwin* (1887). Public domain.

Figure 2.2. DNA rungs. Image by Joseph Condeelis.

Figure 2.3. The ribosome. Image by Joseph Condeelis.

Figure 2.4. Inside the ribosome tunnel. Image by Tim Doherty, Illustra Media.

Figure 2.5. Kinesin molecular machine. "3D Rendered Illustration of a Motor Protein." Image by Sebastian Kaulitzki, Adobe Stock. Standard license.

Figure 3.1. Stephen Quake. Photograph by Christopher Michel, 2024, Wikimedia Commons. CC-BY-SA 4.0 International license.

Figure 3.2. The nucleus. Image by Joseph Condeelis.

Figure 3.3. A histone core spool with two windings of DNA. Image by Joseph Condeelis.

Figure 3.4. DNA with methyl tags attached to certain C-letters. Image by Joseph Condeelis.

Figure 4.1. Jonathan Wells. Photograph by Laszlo Bencze.

Figure 4.2. Glycans branching out like trees from a cell. Image by Pixel Painters.

Figure 4.3. The spherical centrosome. Image by Joseph Condeelis.

Figure 5.1. Mice. "Agouti Mice." Photograph by Randy Jirtle and Dana Dolinoy, 2007, Wikimedia Commons. CC-A 3.0 Unported license.

Figure 5.2. Some nutrients, and foods, helpful for maintaining a healthy epigenome. Table by Thomas Woodward.

Figures 6.1. The "INF-β" enhanceosome complex. Image by David Goodsell, RCSB PDB Molecule of the Month, February 2010, http://doi.org/10.2210/rcsb_pdb/mom_2010_2.

Figure 7.1. James Watson. Photograph by staff photographer, *New York World-Telegram and Sun*, prior to 1968. Public domain.

Figure 7.2. Michael Behe. Photograph by Chris Morgan.

Figure 7.3. The bacterial flagellum. Image by Tim Doherty, Illustra Media.

Figure 8.1. Sagrada Família church in Barcelona, Spain. Photograph by C. Messier, August 24, 2017, Wikimedia Commons. The cranes have been digitally removed. CCA-SA 4.0 license.

Figure 8.2. Byron Johnson. Photograph by Baylor University.

ABOUT THE AUTHORS

Thomas Woodward is a Professor Emeritus at Trinity College of Florida and a Fellow of the Center for Science & Culture. A Princeton graduate in history and Latin American Studies, he completed his PhD in the Rhetoric of Science at the University of South Florida. His dissertation on the history of intelligent design positioned him as an early historian of intelligent design. His books include the 2002 book *Darwinism Under the Microscope* (with James P. Gills), *Darwin Strikes Back: Defending the Science of Intelligent Design* (2006), and *Doubts About Darwin: A Rhetorical History of Intelligent Design*, named a *Christianity Today* Book of the Year in 2004. Dr. Woodward is founder and senior lecturer of the C. S. Lewis Society (renamed Apologetics, Inc.), a non-profit organization that explores questions related to science and the Christian faith. He is also founder and president of the non-profit DNA and Beyond, which produces physical models useful in teaching the complexity of DNA and molecular biology, including epigenetics. Woodward has lectured on intelligent design in thirty-eight countries, and he launched and co-hosted *The Universe Next Door* podcast.

James P. Gills trained at Johns Hopkins as a resident physician in the 1960s and is founder and director of St. Luke's Cataract and Laser Institute in Tarpon Springs, Florida. He was the first ophthalmologist in the United States to dedicate his practice to the treatment of cataracts via the use of intraocular lens implants and, according to his page at Johns Hopkins University, has since performed more cataract and lens implant surgeries than anyone else in the world. A trailblazer in the field, he developed and refined intraocular lens implant techniques and procedures that continue to guide cataract surgery methods today. He also applied new uses for medications that have improved the post-operative recovery process, establishing standards embraced by cataract surgeons globally. Dr. Gills is the author of many books, including *Darwinism Under the Microscope* (with Thomas Woodward), *God's RX for Depression and Anxiety* and *God's Prescription for Healing*. Gills has provided an exceptional amount of charitable care to patients through his cataract and laser institute. The James P. Gills Professorship in Ophthalmology at Johns Hopkins is named in his honor.

INDEX